T0332927

The Dryinidae and Embolemidae (Hymenoptera: Chrysidoidea) of Fennoscandia and Denmark

FAUNA ENTOMOLOGICA SCANDINAVICA
Volume 30 1994

The Dryinidae and Embolemidae (Hymenoptera: Chrysidoidea) of Fennoscandia and Denmark

by

M. Olmi

E.J. Brill
Leiden · New York · Köln

Study supported by a grant of the Italian Ministry of the University and Scientific and Technologic Research, 40%.

The paper in this book meets the guidelines for permanence and durability of the Committee on Production Guidelines for Book Longevity of the Council on Library Resources.

PRINTED IN THE NETHERLANDS

Editor-in-chief: N.P. Kristensen
Desk editor: V. Michelsen

ISBN 90 04 10224 8
ISSN 0106-8377

Library of Congress Cataloging-in-Publication Data

Olmi, Massimo.
The Dryinidae and Embolemidae (Hymenoptera: Chrysidoidea) of Fennoscandia and Denmark / by M. Olmi.
p. cm. — (Fauna Entomologica Scandinavica ; v. 30)
Includes bibliographical references (p.) and index.
ISBN 9004102248 (cloth : alk. paper)
1. Dryinidae—Scandinavia. 2. Embolemidae—Scandinavia.
3. Dryinidae—Scandinavia—Classification. 4. Embolemidae—Scandinavia—Classification.
I. Title. II. Series.
QL568.D7046 1994
595.79—dc20
94-38823
CIP

Die Deutsche Bibliothek - CIP-Einheitsaufnahme

Olmi, M.:
The Dryinidae and Embolemidae (Hymenoptera : Chrysidoidea) of Fennoscandia and Denmark / by M. Olmi. — Leiden ; New York ; Köln : Brill 1994
(Fauna Entomologica Scandinavica ; Vol. 30)
ISBN 90-04-10224-8
NE : GT

Cover illustration: Female of *Anteon pubicorne* (Dalman)

Author's address:
Prof. Massimo Olmi,
Dipartimento di Protezione delle piante,
Universitá della Tuscia,
Via S. Camillo de Lellis,
01100 VITERBO, Italia

Contents

38 colour plates are given between p. 88 and 89.

Introduction

Our knowledge of the Dryinidae and Embolemidae of Fennoscandia and Denmark began with S.J. Ljungh (1810) and the description of *Gonatopus formicarius*, a new genus and species from Sweden.

Linnaeus (1767) had previously described a dryinid (*Sphex* (= *Dryinus*) *collaris*), but it was from Spain (Day, 1979).

After Ljungh and his species, the most important contributions to knowledge of the Fennoscandian Dryinidae during the nineteenth century were by J.W. Dalman (1818, 1823). He described a number of species from Sweden, eight of which are still valid. During this period, Thunberg (1827) described *Gelis clavipes*, a new species from Sweden, which has always been considered to be an ichneumonid but which in fact is a good species of the dryinid genus *Gonatopus*. Some years later, C.G. Thomson (1860) described the females of five species of Dryinidae from Sweden; none is valid today (Olmi, 1977). The nineteenth century closed with the catalogue of Norwegian Hymenoptera by E. Strand (1898), which contained the first record of Norwegian Dryinidae with a single species from that country (*Dryinus* (= *Lonchodryinus*) *ruficornis* Dalman).

The early twentieth century saw the first paper on Finnish Dryinidae, by J. Sahlberg (1910). He listed five species from Finland, none of which is valid today.

Elsewhere in Europe, the first 15 years of the new century were marked by the important work on Dryinidae, Bethylidae and Embolemidae by the Abbé J.-J. Kieffer. In collaboration with T.A. Marshall, he began with a Palaearctic revision for André's series 'Species des Hyménoptères d'Europe et d'Algérie' (1904-1906), and his work culminated in a world monograph (1914). Between and before these two books, a number of smaller papers (among which was a catalogue for the series 'Genera Insectorum' (1907)) contributed substantially towards a fuller knowledge of the Dryinidae and Embolemidae of the world and also included species from Fennoscandia and Denmark. Kieffer's work is now of little more than historical significance because his descriptions are hardly reliable and the classification has been completely transformed (Olmi, 1984, 1993).

The first half of the twentieth century in Finland produced work by R. Forsius (1925), W. Hellén (1919a, b, 1930, 1935, 1946, 1953) and H. Lindberg (1950). In his catalogue of 1935 (with R. Forsius) Hellén listed 2 species of Embolemidae and 45 species of Dryinidae from Finland. Hellén's most significant paper, however, was his revised catalogue of 1953, published after O.W. Richards' (1939) major revision of the Dryinidae and Embolemidae.

The paper by Richards can be considered to be the first modern publication on the Dryinidae and Embolemidae after the era dominated by Kieffer's unreliable papers. Not unnaturally, Hellén (1953) took advantage of the new classification proposed by Richards.

Richards' 1939 paper was also an important contribution to knowledge of the Dryinidae and Embolemidae of Fennoscandia and Denmark because there are many references to species and localities from these countries. In addition, Richards was the first to draw the attention of European entomologists to the fundamental papers on Dryinidae by R.C.L. Perkins. The work of Perkins was virtually unknown in Europe because his most important papers (1903, 1905, 1906a, 1906b, 1906c, 1907, 1912) were published in an almost unknown review, 'Report of work of the Experiment Station of the Hawaiian Sugar Planters' Association'. Few of the European entomologists working on Dryinidae in the first half of the twentieth century were aware of the fundamental systematic information contained in Perkins' papers.

The paper by Lindberg (1950) was one of the first to deal with the biology of the Dryinidae and was the first paper on the biology of Scandinavian Dryinidae. It can be regarded as a modern work, and is still cited in almost all papers on dryinid biology.

The first half of the twentieth century also saw the publication of papers on the ecology of dryinids, such as that by R. Krogerus (1932) on the arthropod populations of Finland's coastal drifting sands. Ten species of Dryinidae were listed from this interesting environment.

Research in Finland after World War II gave rise to papers dealing with the biology of several species of planthoppers of economic importance which contained biological data on dryinid parasitoids. They are the papers on *Megadelphax sordidulus*

(Stål), *Unkanodes excisa* (Melichar), *Dicranotropis hamata* (Boheman) and *Javesella pellucida* (F.) by Raatikainen (1960, 1961, 1967, 1970), Raatikainen and Vasarainen (1964) and Heikinheimo (1957). A paper by Kontkanen (1950) dealt with the parasitoids of leafhoppers in North Karelia, and included much biological data on dryinids.

In Sweden the only paper on dryinids during this period was by A. Jansson (1950a), with the description of a new species, whilst the Hungarian L. Moczar (1967) revised Dalman's types. So far as the Embolemidae are concerned, A. Jansson (1950b) and subsequently K.-J. Hedqvist (1975) listed Swedish localities for the single Scandinavian species.

More recently, N.G. Ponomarenko (1978) and V.A. Tryapitsyn (1978) have published keys to the Dryinidae and Embolemidae of the European part of the former Soviet Union, which were partly applicable to the Scandinavian species. However, these keys were based on old taxonomic concepts and have already been superseded by the publication in recent years of the revisions by H. Hilpert (1989) of the Embolemidae and by M. Olmi (1977, 1984, 1989, 1993) and M. Olmi and I. Currado (1977) of the Dryinidae. In fact, Hilpert synonymised the two species of Embolemids cited by Tryapitsyn, and Olmi has completely transformed the old systematics, basing his studies on a revision of the world species.

More recently, faunistic records of Dryinidae and Embolemidae have been published by G.E. Nilsson (1986, 1988, 1991) for Sweden, M.A. Jervis (1986) for Finland, L. Greve and E. Hauge (1989) and L.O. Hansen and M. Olmi (1994) for Norway. Olmi (1994) has revised the world species of the Embolemidae.

The colour illustrations for the present book have been prepared in part by Miss Monica Cirillo and in part by Mr Luca Palermo. The other drawings have been prepared by the author and by Mr Nicolò Falchi. The male genitalia of all species have been figured as well as the female chelae.

Synonyms are given only if they have been used in the older Danish and Fennoscandian literature, as well as in the most important revisions of Dryinidae and Embolemidae.

For each species, the distribution within and outside Denmark and Fennoscandia is given. Information on localities in Denmark was supplied by Dr Børge Petersen; in Sweden by Dr Karl-Johan Hedqvist, Dr Roy Danielsson and Dr Göran Nilsson; in Norway by Dr Lita Greve and Dr Lars Ove Hansen; in Finland by Dr Anders Albrecht. The author has personally identified all the specimens recorded in the present book, and has also checked all the localities with the assistance of the colleagues cited above. Virtually all the specimens of Dryinidae and Embolemidae from Denmark and Fennoscandia kept in world museums have been examined and checked, including all type material. Some of the collection records are the results of personal rearings and field collections by the author.

Host records and names have been taken from the literature, except for data resulting from rearings by the author; in these cases, the host identifications have been made by Dr R. Remane, Dr W. della Giustina, Dr J. Dlabola, Dr V. D'Urso and the late Dr Carlo Vidano.

Information on the biology of the species has been completed from the literature or from personal observations by the author. The bibliography has been made as complete as possible.

For host terminology, the author has followed Ossiannilsson (1978, 1981, 1983), except for the names of the immature stages of Auchenorrhyncha. I believe that it is better to distinguish the juvenile stages of Exopterygota as 'nymphs' (= larvae, sensu Ossiannilsson, 1978, 1981, 1983), as is done by most modern authors (see, for example, 'The Insects of Australia', Melbourne Univ. Press, 1991, but also Wilson and Claridge, 1991).

Acknowledgements

For the generous loan of material, the author is greatly indebted to Dr Anders Albrecht, Zoological Museum, Helsinki; Dr Roy Danielsson, Zoological Museum, Lund; Dr Michael Day, Natural History Museum, London; Dr Karl-Johan Hedqvist, Natural History Museum, Stockholm; Dr Lita Greve, Zoological Museum, Bergen; Dr Lars Ove Hansen, Drammen, Norway; Dr Frank Koch, Museum für Naturkunde der Humboldt-Universität, Berlin; Dr Ole Lomholdt, Zoological Museum, Copenhagen; Dr Lubomir Masner, Biosystematics Research Centre, Ottawa; Dr Arnold Menke, National Museum of Natural History, Washington; Dr Anders Nilsson, Department of Animal Ecology, Umeå; Dr Göran Nilsson, Department of Zoophysiology, Uppsala; Dr Jenö Papp, Hungarian Natural History Museum, Budapest; Dr Børge Petersen, Zoological Museum, Copenhagen; Dr

Nadezdha Ponomarenko, Institute of Animal Morphology and Ecology, Moscow; Dr Martin Schwarz, Institut für Zoologie, Salzburg; the late Dr Henry Townes, American Entomological Institute, formerly in Ann Arbor; Dr Lars Wallin, Zoological Museum, Uppsala.

For identifications of the Auchenorrhyncha hosts, the author is particularly indebted to Dr Reinhard Remane, Zoological Institute of the Philipps-Universität, Marburg/Lahn, Germany; Dr William della Giustina, Zoological Station of INRA, Versailles, France; Dr Vera D'Urso, Department of Animal Biology, Catania; Dr Jirí Dlabola, Zoological Museum, Prague; and the late Dr Carlo Vidano, Institute of Agricultural Entomology, Turin.

The author thanks Miss Monica Cirillo and Mr Luca Palermo, for their excellent colour illustrations, and Mr Nicolò Falchi, for some line drawings. Thanks are also due to Mr Massimo Vollaro for his colour photographs (Plates 1-4).

Survey of the superfamily Chrysidoidea

The Embolemidae and Dryinidae are here treated as separate families of the superfamily Chrysidoidea. This superfamily, together with the Vespoidea and Apoidea, belongs to the hymenopterous division Aculeata (Gauld and Bolton, 1988). The two divisions Aculeata and Parasitica make up the hymenopterous suborder Apocrita.

The Chrysidoidea have previously been known as the Bethyloidea. However, Day (1977) has shown that the name Chrysidoidea has priority over Bethyloidea.

According to Gauld and Bolton (1988), Finnamore and Brothers (1993), and Naumann (1991), the superfamily Chrysidoidea contains seven families which together form a holophyletic group within the Aculeata. There are four families present in Fennoscandia and Denmark: Embolemidae, Dryinidae, Bethylidae and Chrysididae. The three families not represented are the Sclerogibbidae (Pantropical, but reaching South Europe), Plumariidae (South America and South Africa) and Scolebythidae (South Africa, Madagascar, South America, Australia).

The Chrysidoidea are among the smallest Aculeata, with sexual dimorphism often well developed. They can be fully winged or brachypterous or apterous. The antennae are 10-, 12- or 13-segmented. The wing venation is reduced, usually with 1-3 closed cells in the fore wing (rarely up to 8 in Plumariidae) and 1 or no closed cells in the hind wing (rarely up to 3 in Plumariidae). The loss of the jugal lobe from the hind wing is an autapomorphy for the Chrysidoidea. The ovipositor is concealed when at rest and is usually modified as a sting. An articulation within the 2nd gonocoxa (= second valvifer, sensu Carpenter, 1986) is another autapomorphy for the superfamily.

The Chrysidoidea are parasitoids or predators of Lepidoptera, Coleoptera, Embioptera, Homoptera Auchenorrhyncha and Cheleutoptera (= Phasmatodea). Some species are cleptoparasites in the nests of other Aculeata. The world fauna is estimated to be around 16,000 species (Finnamore and Brothers, 1993).

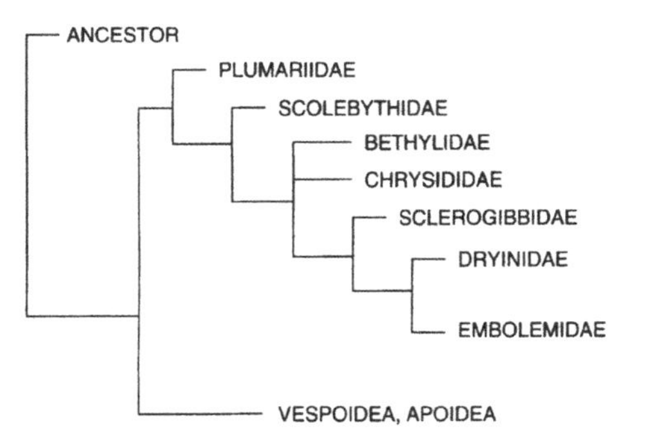

Tab. 1. Phylogeny of Chrysidoidea, from Brothers and Carpenter (1993).

Classification of the Chrysidoidea

The following key will separate the families of Chrysidoidea represented in Fennoscandia and Denmark (modified from Finnamore and Brothers, 1993):

1 Antennae 10-segmented 2
– Antennae 12- or 13-segmented 3
2 Antennae inserted far from mouth, on a frontal prominence (Fig. 1) . Embolemidae
– Antennae inserted near mouth, not on a frontal prominence (Fig. 2) ..... Dryinidae
3 Metasoma with 6-7 exposed terga; pronotum usually touching tegula; head usually prognathous .... Bethylidae
– Metasoma with 5 or fewer exposed terga,

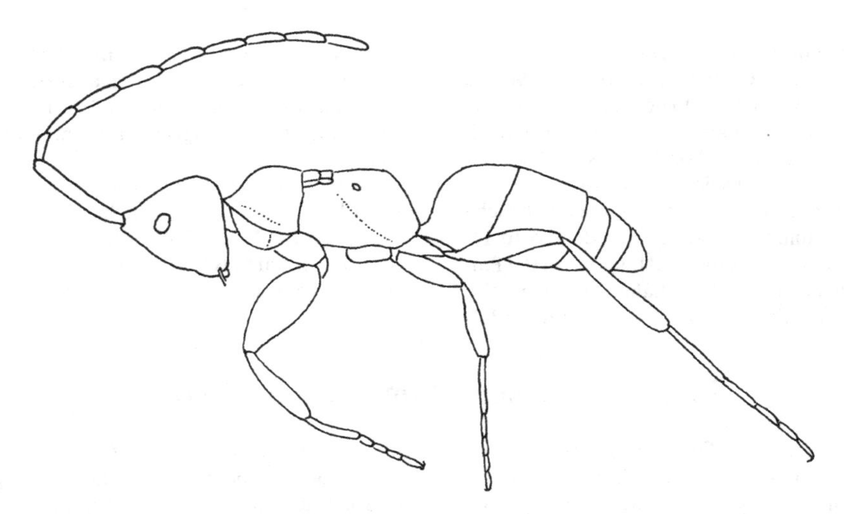

Fig. 1. Female of *Embolemus ruddii* Westwood in lateral view.

rarely with indications of a 6th; pronotum usually separated from tegula; head usually hypognathous .... Chrysididae

According to Brothers and Carpenter (1993), the Chrysidoidea is a holophyletic group with the Plumariidae as its most basal taxon. The Scolebythidae is the next most basal taxon, and the Bethylidae + Chrysididae is the sister-group of the Sclerogibbidae + (Dryinidae + Embolemidae). The Dryinidae is the sister-group of the Embolemidae. The other two superfamilies of the Aculeata, Apoidea and Vespoidea, together form a second holophyletic group.

The cladogram (Tab. 1) proposed by Brothers and Carpenter (1993) shows the possible phylogenetic relationships among the Chrysidoidea.

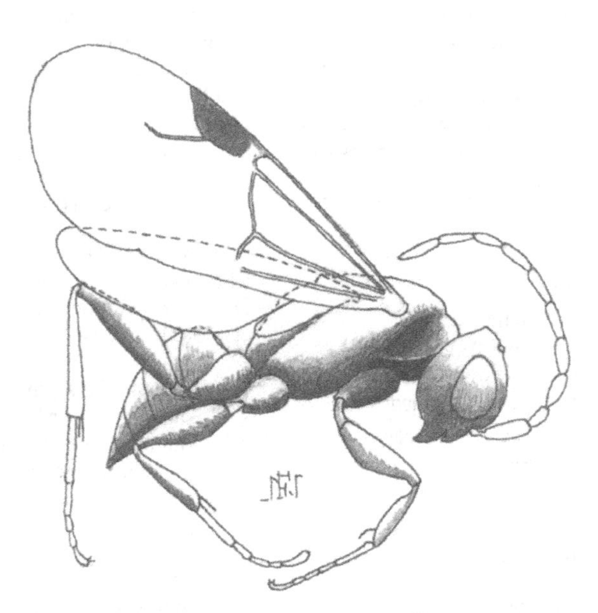

Fig. 2. Male of *Anteon tripartitum* Kieffer in lateral view.

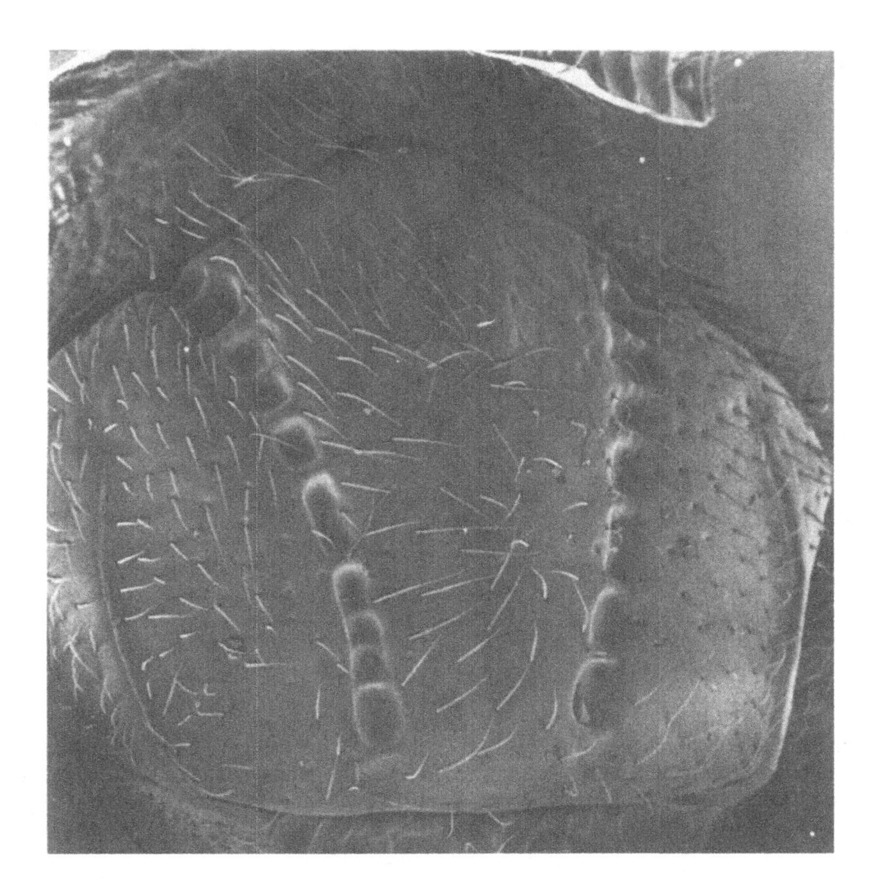

Fig. 3. Punctate sculpture.

Morphology and diagnostic characters of Dryinidae and Embolemidae

Adults

The following account of the morphology of the Dryinidae and Embolemidae is based on the dryinid world revision by Olmi (1984), modified according to Gauld and Bolton (1988). It does not give a complete morphological description of the Dryinidae and Embolemidae, but only elucidates the terminology used in this paper.

Sculpture (Figs 3-6)

The surface of the body is usually sculptured in a variety of ways. The main terms used in describing the sculpture are as follows: without sculpture (smooth); punctate (Fig. 3); granulate (Fig. 4); reticulate rugose (Fig. 5); rugose; striate (Fig. 6) (see also Olmi, 1984, and Eady, 1968, for further details).

Head and its appendages (Figs 1, 7-11)

The head is characteristically hypognathous (= orthognathous), with a ventrally-directed mouth. The compound eyes (Fig. 7 A) are situated laterally on the head. Usually numerous hairs arise from between the facets. The vertex (Fig. 7 B) is the top of the head, the area between the compound eyes; it usually has three ocelli (Fig. 7 C). The ocelli are absent only in the females of Apodryininae and Plesiodryininae, exotic subfamilies of Dryinidae, and in micropterous females of Embolemidae. The face (Fig. 7 D) is the area between the mouth margin and the anterior ocellus. It is made up of the clypeus (Fig. 7 E), lower face and upper face. The antennae are articulated in sockets (toruli) (Fig. 7

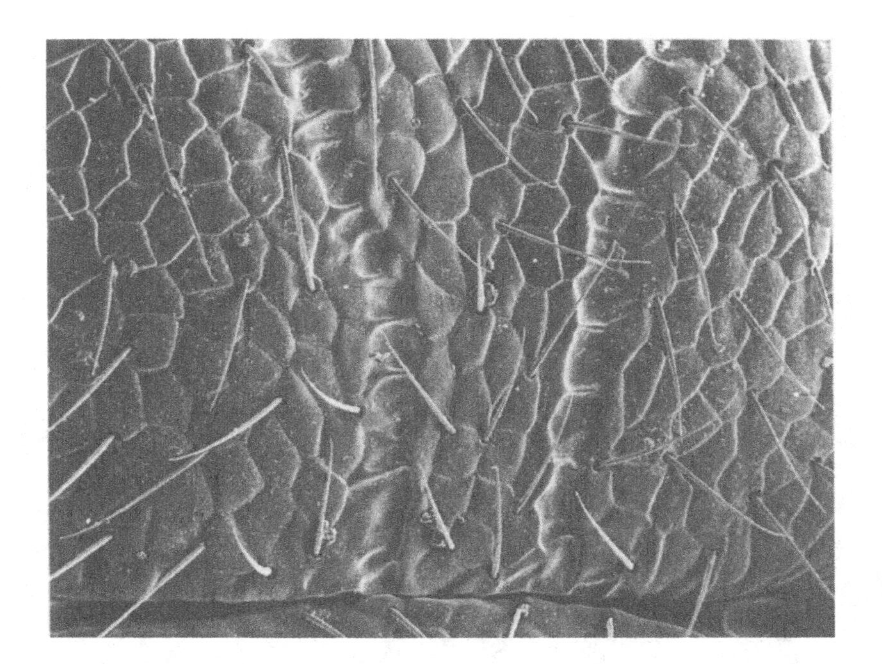

Fig. 4. Granulate sculpure.

F), situated in the face near the clypeus (in Dryinidae) or very far from it (in Embolemidae). In Embolemidae the antennae are articulated on a strong frontal process (Fig. 1). Laterally and dorsally the clypeus is defined by the curved epistomal suture (= clypeal suture) (Fig. 7 G). The frons (Fig. 7 H) is

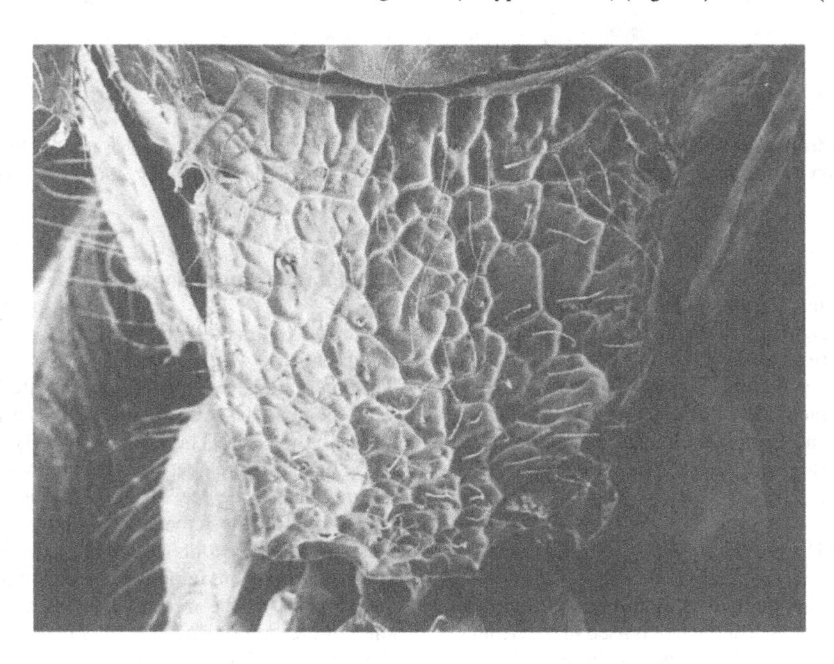

Fig. 5. Reticulate rugose sculpture.

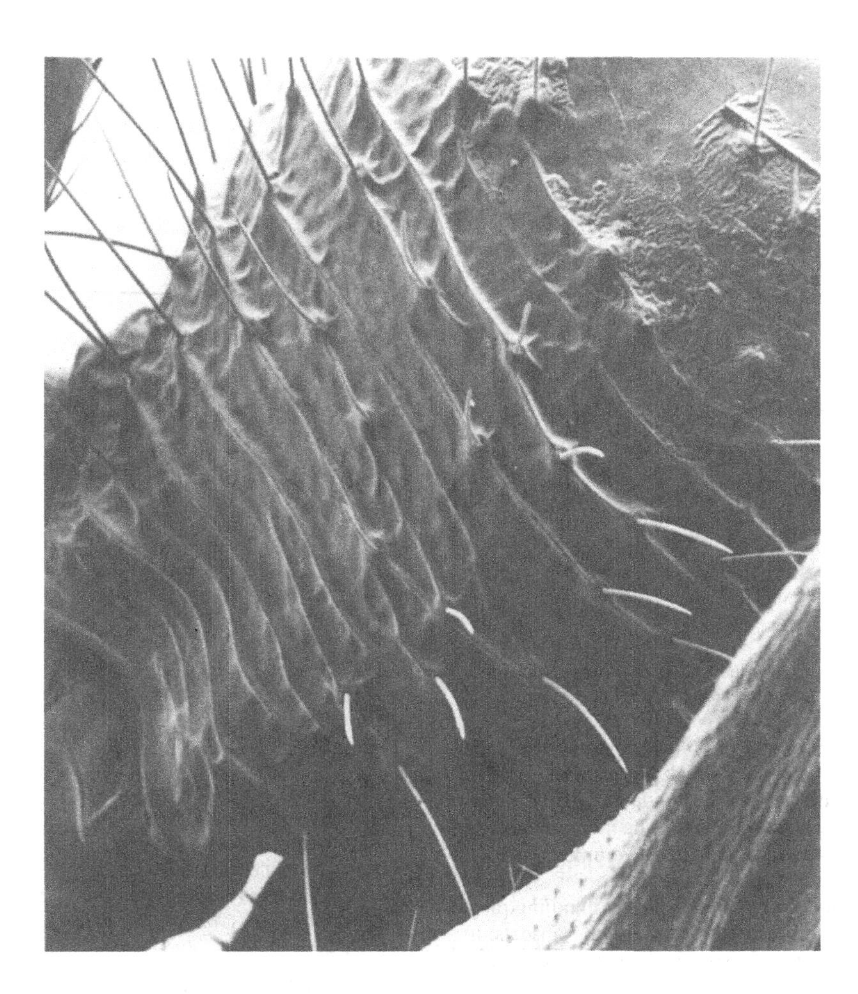

Fig. 6. Striate surface.

the area between the antennal sockets and the anterior ocellus. It is often divided by a central longitudinal keel (frontal line) (Fig. 8 A). In many males of Dryinidae the frons is more or less excavated to receive the basal part of the antennae; these excavations are the antennal scrobes. The vertex is limited posteriorly by the occipital carina (Fig. 9 A); this keel can be absent in many Dryinidae (in most Gonatopodinae, and a few Dryininae, Apodryininae and Plesiodryininae). The ocelli vary greatly in position; the distance from the outer edge of a lateral ocellus to the compound eye is the ocular-ocellar line (OOL); the distance between the inner edges of the two lateral ocelli is the postocellar line (POL); the distance from the inner edge of the anterior ocellus to the inner edge of a lateral ocellus is the ocellar line (OL); the distance of the lateral ocelli from the occipital carina is the ocellar-occipital line (OPL). The foramen magnum is an approximately circular opening in the skeleton of the back of the head through which the internal organs pass from the head to the thorax. The occiput is the area between the occipital carina and the foramen magnum. If the occipital carina is absent the vertex is not marked off from the occiput. The ventral part of the occipital carina is often called genal carina. The area between the compound eyes and the genal carina is the gena (Fig. 10 A). The upper part of the gena is called temple (Fig. 9 B). Viewed dorsally, the distance from the posterior edge of a compound eye to the genal carina is the ocular-occipital line (TL). If the genal carina is absent, the occiput and gena are continuous. The region between the base of a mandible

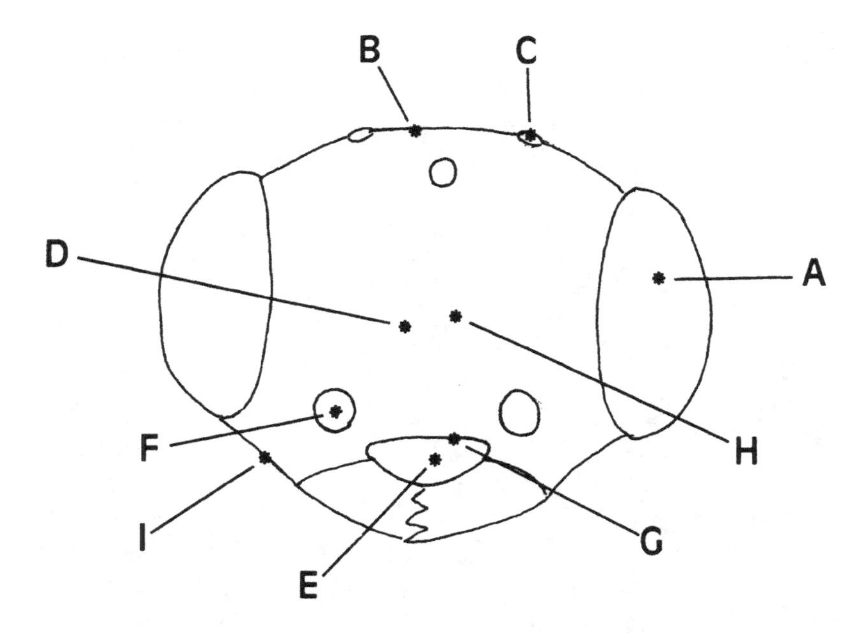

Fig. 7. Head of dryinid in frontal view. A: compound eyes; B: vertex; C: ocellus; D: face; E: clypeus; F: torulus; G: epistomal suture; H: frons; I: malar space.

and the ventral margin of a compound eye is the malar space (Fig. 7 I), which is part of the gena.

The mouth-parts of Dryinidae and Embolemidae are simply mandibulate. The mandibles stand apart from the labio-maxillary complex. Perhaps because they are used for such purposes as breaking open the cocoon and killing or handling prey, they have not been reduced as in other insects. In Dryinidae and Embolemidae, however, the number of mandibular teeth may be greatly reduced and may vary from 1 to 4. The maxillae and labium are firmly united to each other by membranes and form a single labio-maxillary complex. Each of the two maxillae consists of a basal piece, the cardo, and a distal lobe, the stipes. Externally the stipes bears the maxillary palp, which primitively is composed of 6 segments, but frequently has fewer (2-5). Internally the stipes bears two lobes, a more proximal one, the lacinia, and a more distal one, the galea. The labium consists of a submentum and a prementum. Externally, the distal end of the prementum bears the two labial palpi, which consist typically of three segments though the number may be reduced (1-2). In the Dryinidae and Embolemidae of Fennoscandia and Denmark the palp formula may be summarized as follows (first number = maxillary palp-segments; second number = labial palp-segments):

Dryinidae

Aphelopinae (*Aphelopus*): 5/2
Anteoninae (*Lonchodryinus*): 6/3
Anteoninae (*Anteon*): 6/3
Dryininae (*Dryinus*): 6/3
Gonatopodinae (*Haplogonatopus*):2/1
Gonatopodinae (*Gonatopus*): 6/3, 5/3, 4/3, 5/2, 4/2, 3/2, 2/2

Embolemidae (*Embolemus*): female: 4/2; male: 6/3

The maxillary and labial palpi are usually borne on special lobes, the palpigers, of the stipes and prementum respectively. These lobes may look like a basal segment of the palp. In this paper, and also according to Olmi (1984), the first small basal segment of the palpi is considered a true joint. Except when the palp formula is 2/1, the maxillary palpi are elbowed after the two basal segments. In the Dryinidae, the maxillary and labial palpi have been used extensively in the classification of the Gonatopodinae for generic differentiation. According to Olmi (1984), however, it appears that the number of palpal segments may vary within species and within genera. Moreover, the last two segments of the maxillary palpi may be fused.

The antennae of Dryinidae are filiform, not thickened distally in males, slightly thickened dis-

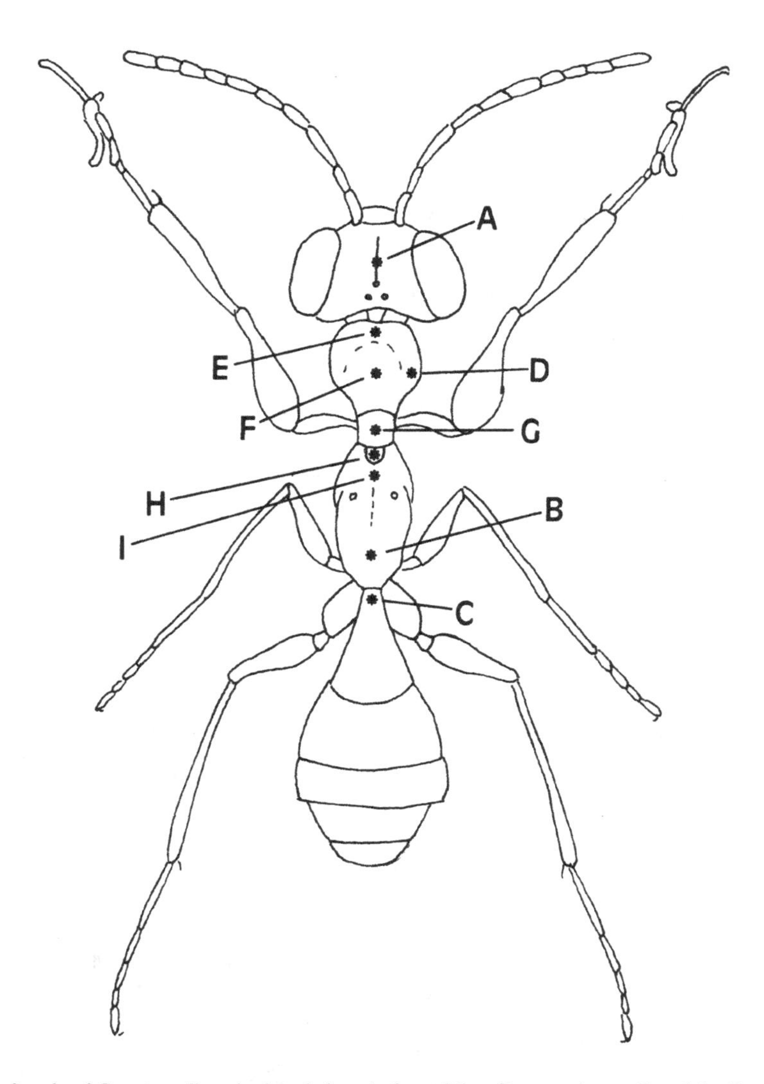

Fig. 8. Apterous female of Gonatopodinae in dorsal view. A: frontal line; B: propodeum; C: petiole; D: pronotum; E: pronotal anterior collar; F: pronotal disc; G: scutum; H: scutellum; I: metanotum.

tally in females. The females rarely have antennae not thickened distally and the males have antennae pectinate. In the Embolemidae the antennae are also filiform, thickened or not thickened distally. The antennae are occasionally elbowed (geniculate) after the two basal segments in both Dryinidae and Embolemidae. In both families the antennae are composed of 10 segments. The first segment is the scape, occasionally very long; the second segment is the pedicel, usually short; the remainder of the antenna forms the flagellum. The surface of the antennae is densely covered with sense-organs of various types. These are sometimes of taxonomic importance, especially when large enough to be visible without great magnification. For instance, in the exotic dryinid genus *Thaumatodryinus*, the females have tufts of long hairs on antennal segments 5-10. In Apodryininae, Plesiodryininae, the majority of Dryininae and many Gonatopodinae, the females have longitudinal groove-like structures, the rhinaria (Fig. 11), on the distal segments. The number of segments with rhinaria varies within genus and subfamily (2-5), but is constant within species. Apparently rhinaria are not visible in females of species parasitizing the Cicadomorpha; they are present only in females

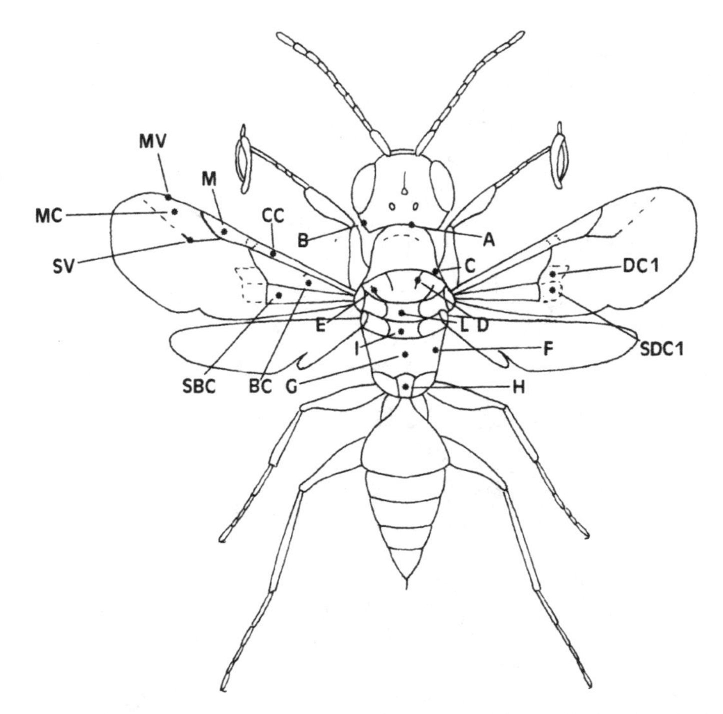

Fig. 9. Winged female of Anteoninae. A: occipital carina; B: temple; BC: basal cell; C: pronotal tubercle; CC: costal cell; D: notauli; DC1: first discal cell; E: parapsidal lines; F: propodeum; G: propodeal dorsal region; H: propodeal posterior area; I: metanotum; L: scutellum; M: pterostigma; MC: marginal cell; MV: marginal vein; SBC: subbasal cell; SDC1: first subdiscal cell; SV: stigmal vein.

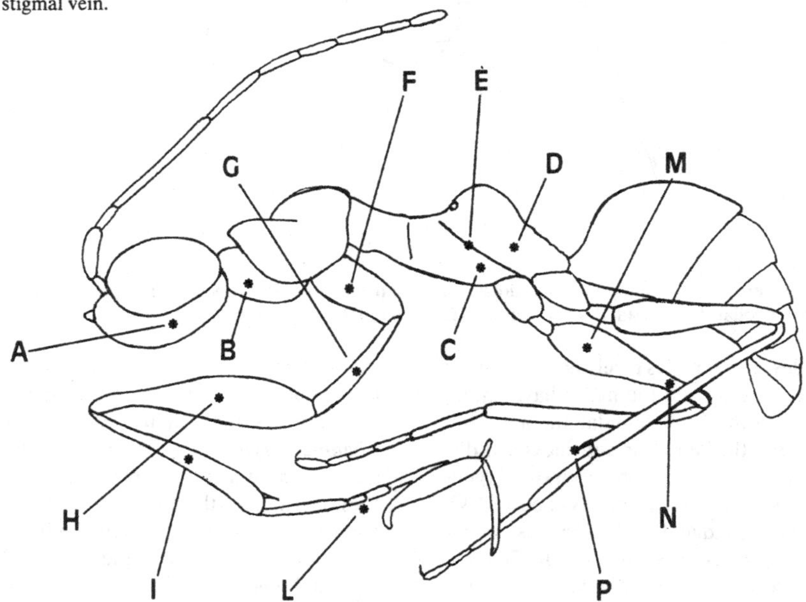

Fig. 10. Apterous female of Gonatopodinae in lateral view. A: gena; B: propleura; C: mesopleura; D: metapleura; E: meso-metapleural suture; F: coxa; G: trochanter; H: femur; I: tibia; L: tarsus; M: club of femur; N: stalk of femur; P: tibial spur.

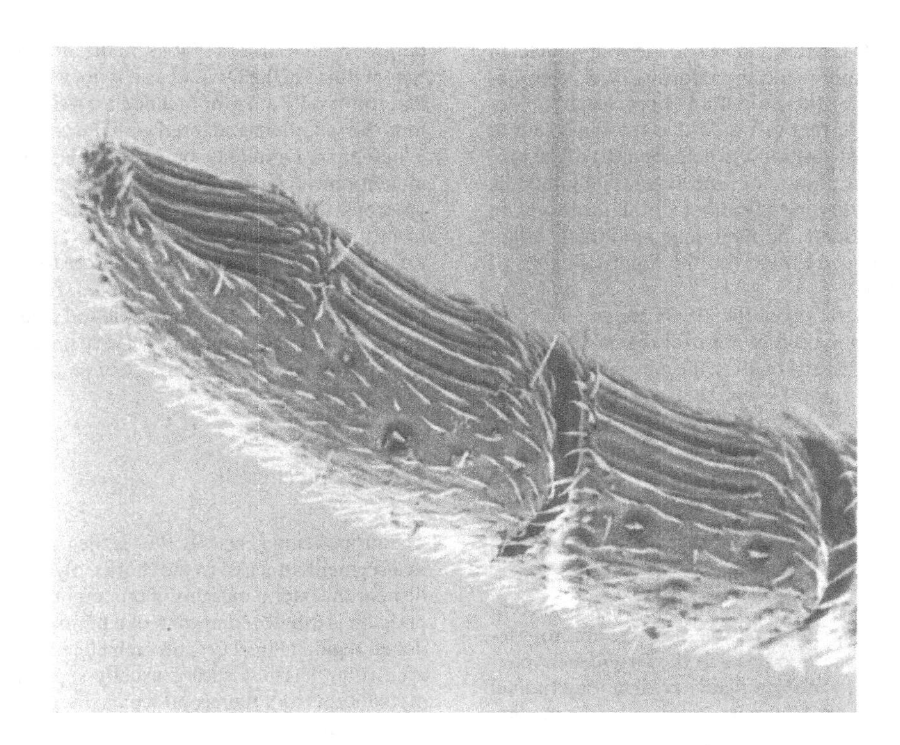

Fig. 11. Antennal rhinaria.

parasitizing the Fulgoromorpha. The genus *Thaumatodryinus* is the only exception, as its hosts are Flatidae (Fulgoromorpha) and its females do not have rhinaria. However, they have tufts of long hairs, probably playing the same rôle. In females of Embolemidae there are no rhinaria or tufts of long hairs although they parasitize Achilidae (Fulgoromorpha). Rhinaria and antennal tufts are certainly apomorphies, whereas embolemids follow more plesiomorphic strategies for host recognition.

Thorax (Figs 8-10)

The thorax consists of three segments, prothorax, mesothorax and metathorax. The first abdominal segment (propodeum (Fig. 8 B)) is firmly joined to the thorax and is connected to the rest of the abdomen by a narrow petiole (Fig. 8 C). The thorax of Dryinidae and Embolemidae (as in all the Apocrita) thus appears to consist of four segments. This structure is called mesosoma (= alitrunk). In Apodryininae and Plesiodryininae most of the mesosoma segments are fused.

The prothorax consists of a dorsal sclerite, the pronotum (Fig. 8 D), laterally extended to form the largest prothoracic region. Ventrally there are the two propleura (Fig. 10 B), between which there is a small prosternum. The propleura form almost all the ventral part of the prothorax and are also called propectus (sensu Olmi, 1984). They meet along the median ventral line. Anteriorly the propleura support the head. In most Embolemidae and, among the Dryinidae, in males and in the less evolved females, the propleura are not visible from above, whereas in the more specialized dryinid females they are partly visible in front of the pronotum, behind the head (Fig. 10 B). In Dryinidae the pronotum is commonly divided by an anterior transverse furrow into an anterior region (anterior collar (Fig. 8 E)) and a more posterior part (disc (Fig. 8 F)). Mostly in the subfamily Dryininae, a posterior transverse furrow is also visible, defining a posterior collar. No transverse furrow is visible in some females of Gonatopodinae. In the males of Dryinidae, as well as the females of Aphelopinae, Biaphelopinae and Conganteoninae, the pronotum is very reduced dorsally and almost invisible from above. In more specialized dryinid females the pronotum is produced backwards into pronotal tubercles (Fig. 9 C) directed towards the tegulae. The pronotal tubercles are only absent in females of Gonatopodinae and Plesiodryininae. In females of more specialized forms (Dryininae, Gonatopodinae) the pronotum is remarkably elongated, and further forward the extension is caused by the dorsal position and development of the propleura. The pronotum itself is unusually mobile. These peculiar modifications of the prothorax in Gonatopodinae and Dryininae are largely adaptive, being correlated with the raptorial habits of the females.

The mesothorax of the Dryinidae and Embolemidae is composed of a dorsal region, the mesonotum, two lateral regions, the mesopleura (Fig. 10 C), and a ventral area, the mesosternum. The mesonotum is divided into two sclerites, the scutum (situated anteriorly) (Fig. 8 G) and the scutellum (situated posteriorly) (Figs 8 H, 9 L). From the front margin of the scutum two sulci, the notauli (= notaulices, sensu Olmi, 1984) (Fig. 9 D), often converge posteriorly. The notauli may be deep and strong and meet at about the centre of the posterior margin of the scutum; or they may be incomplete or completely invisible, as in the apterous females of Gonatopodinae. In the Dryinidae Aphelopinae and Embolemidae a median longitudinal groove, the median mesoscutal sulcus (= median scutal line, sensu Olmi, 1984), not infrequently extends posteriorly from the anterior margin of the scutum, bisecting the scutum. It may be complete or reduced to a short line. According to Gauld and Bolton (1988) this groove is an archaic feature of Hymenoptera, and can be widely observed in fossils but only in a few extant Apocrita. Numerous Dryinidae have a second pair of scutal marks, the parapsidal lines (= parapsidal furrows, sensu Olmi, 1984)(Fig. 9 E), lateral to the notauli. They are visible only in fully winged forms. The lateral regions of the mesothorax are the mesopleura. They are divided from the metapleura by the meso-metapleural suture (Fig. 10 E). This may be obsolete or strongly marked. In the apterous Gonatopodinae of the Dryinidae, the mesothorax, released from its subservience to the function of flight, has become highly modified and specialised. It forms a long and slender waist connecting the prothorax to the posterior part of the thorax. The dorsum of this waist is formed by the scutum, whereas the scutellum is very small and is situated posteriorly.

The metathorax of Dryinidae and Embolemidae is smaller and less differentiated than the mesothorax. It is composed of a dorsal region, the metanotum (Figs 8 I, 9 I), two lateral areas, the metapleura (Fig. 10 D), and a ventral region, the metasternum. The metanotum is a single central area (= dorsellum, sensu Gauld and Bolton, 1988), defined by lateral depressions. In the apterous Gonatopodinae of the Dryinidae, it is not well defined and forms only a flat or inclined area situated behind the scutellum and fused with the propodeum. Whether in Dryinidae or in Embolemidae, the metanotum is usually smaller in apterous or brachypterous forms. The metapleura are separated from the mesopleura by a meso-metapleural suture, whereas there is usually no suture to separate the metapleura from the propodeum. In apterous Gonatopodinae the meso-metapleural suture may be absent. The metasternum is a very small sclerite situated between the two posterior pairs of coxae.

Propodeum (Fig. 9)

The propodeum (Figs 8 B, 9 F) is the first abdominal segment attached to the thorax. Two spiracles lie near the lateral margins of the tergite. The propodeum is usually composed of a more or less flat dorsal region (Fig. 9 G) and an inclined posterior area (Fig. 9 H). A suture usually separates the metanotum from the propodeum, whereas no suture is usually visible laterally between the metapleura and the propodeum. In the apterous Gonatopodinae of the Dryinidae, the suture between metanotum and propodeum is absent.

Wings (Fig. 9)

The Dryinidae and Embolemidae have two pairs of wings. Some females may be apterous or brachypterous. According to Waloff and Jervis (1987), wing reduction is associated with parasitism of hosts in grassland and other herbage, presumably because the wings hinder movement through such vegetation. The only case of a brachypterous male is that of *Mystrophorus formicaeformis* (Ruthe). Usually the reduction or absence of the wings is accompanied by a corresponding reduction of the mesothoracic and metathoracic segments. In fully-winged forms in Fennoscandia and Denmark, the fore wing has a pterostigma (Fig. 9 M) and 1-4 cells fully enclosed by pigmented veins. Other cells may be visible, but they are surrounded by very evanescent and unpigmented veins. The following cells enclosed by pigmented veins have been identified in the species of Fennoscandia and Denmark: costal (CC), basal (BC) (= median, sensu Olmi, 1984), subbasal (SBC) (= submedian, sensu Olmi, 1984), first discal (DC 1) (Fig. 9). The situation may be summarized as follows:

Embolemidae (Plate N. 6): CC, BC, SBC, DC 1;
Dryinidae Aphelopinae (Plate N. 8): CC;
Dryinidae Anteoninae (Plate N. 9): CC, BC, SBC;
Dryinidae Dryininae (Plate N. 18): CC, BC, SBC;
Dryinidae Gonatopodinae (Plate N. 36): CC, BC, SBC.

In addition to the veins surrounding the enclosed cells, another pigmented vein is usually visible in Dryinidae and Embolemidae: the stigmal vein (= radial vein, sensu Olmi, 1984) (Fig. 9 SV), usually incomplete and not reaching the margin of the wing or reaching the margin as a very evanescent and unpigmented line. The cell surrounded by the pterostigma, stigmal vein and anterior margin of the wing (where there is the marginal vein) is called the marginal cell (Fig. 9 MC) (= radial cell, sensu Olmi, 1984); it is always open, which means that the stigmal vein is incomplete. Another cell not fully enclosed by the pigmented veins is the first subdiscal cell (Fig. 9 SDC 1), visible only in Embolemidae. The hind wing has a marginal vein on the leading edge, but is otherwise without veins or closed cells.

Legs (Figs 10, 12-16)

There is a pair of legs for each thoracic segment. The legs consist of a basal coxa (Fig. 10 F), a short or elongate trochanter (Fig. 10 G), an elongate femur (Fig. 10 H), a slender tibia (Fig. 10 I) and a segmented tarsus (Fig. 10 L). Dryinidae females have relatively elongated fore coxae, whereas in males they are not elongated. The females of the Embolemidae and, among the Dryinidae, the females of the Aphelopinae do not have elongated fore coxae, as in the males. The trochanter is usually a small segment in Embolemidae and in most Dryinidae. In the females of Dryininae and Gonatopodinae, however, the fore trochanter is very elongated (Fig. 10 G). The femur usually consists of a proximal swollen part (club of femur) (Fig. 10 M) and a distal slender part (stalk of femur) (Fig. 10 N). Usually the club of the fore femur is more swollen in females than in males. Only in the Aphelopinae are the legs alike in both sexes. The tibiae are provided with articulated spurs on the ventral side. The three legs may have 0-2 spurs at the distal apex of the tibiae (Fig. 10 P), as follows: Embolemidae (females and males): 1 (fore), 2 (mid), 2 (hind) (in Fennoscandian and Danish species); males of Dryinidae: 1, 1, 2; females of Aphe-

Fig. 12. Chela of *Anteon gaullei* Kieffer. A: arolium; B: 5th segment; C: enlarged claw; D: basal part of 5th segment; E: apical part of 5th segment; F: lamellae.

Fig. 13. Chela of *Haplogonatopus oratorius* (Westwood). A: rudimentary claw; B: arolium.

lopinae, Anteoninae, Dryininae (Dryinidae): 1, 1, 2; females of Gonatopodinae (Dryinidae): 1, 0, 1 (in Fennoscandian and Danish species). The tarsus always consists of five segments. The last tarsal segment has a pair of claws. Between the claws lies the foot-pad or arolium (Fig. 12 A). The front tarsi of most females of Dryinidae are formed into a chela (Fig. 12) for gripping the hosts. The chela is absent in Embolemidae and, among the Dryinidae, in Aphelopinae. The chela is composed of the 5th segment (Fig. 12 B) and an enlarged claw (Fig. 12 C), which is opposable. A second claw, the rudimentary claw (Fig. 13 A), is reduced (in Dryininae and Gonatopodinae) or absent (in Anteoninae (Fig. 12)). The modified fifth segment of the fore tarsus consists of a proximal part (basal part (Fig. 12 D), directly articulated to the fourth segment) and a more or less long, free distal part (apical part (Fig. 12 E)). The inner side of the enlarged claw and the fifth tarsal segment may bear lamellae (Figs 12 F, 14), bristles (Fig. 15), hairs or peg-like hairs. When the chelate dryinid female is walking its chela is raised, whereas the arolium leans against the ground. When the chela is raised and closed the distal apex of the enlarged claw fits into a hook (called in the text 'inner hook') (Fig. 16) situated on the inner side of the 3rd or 2nd tarsal segment.

Modifications in the fore legs are important elements in the evolution of the Dryinidae. The plesiomorphic condition is visible in Embolemidae and Aphelopinae, where fore, mid and hind legs

Fig. 14. Chela of *Anteon gaullei* Kieffer: apical lamellae.

are very similar in both sexes, and where the chelae are absent. This condition is also visible in the males of all the other dryinid subfamilies. The situation changes in the females of the other dryinid subfamilies, where it is possible to assert that evolutionary change, at least as far as the legs are con-

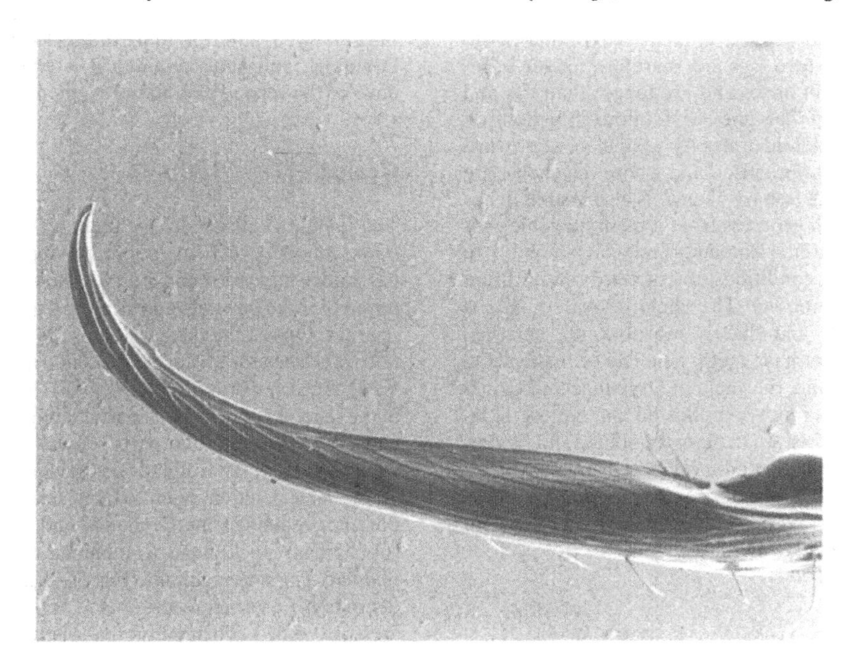

Fig. 15. Enlarged claw of *Gonatopus clavipes* (Thunberg).

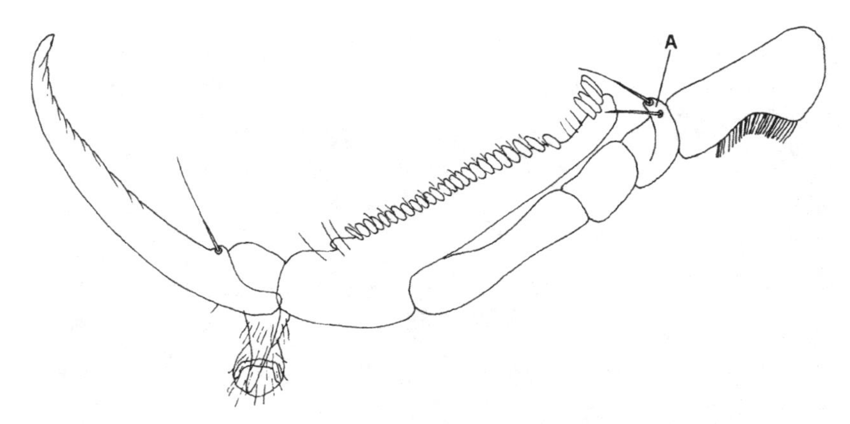

Fig. 16. Fore tarsus of female of *Anteon fulviventre* (Haliday); A: inner hook.

cerned, has been driven by the female sex. Host feeding (which means that adult dryinid females prey on their hosts to obtain food) probably initiated a different evolutionary impulse, together with a better raptorial ability and consequently a better capacity for parasitization. We have observed the appearance of many apomorphies aimed at facilitating host capture. As the dryinid female captures the host with her fore legs, evolutionary change has taken place in the fore legs. The first step is the appearance of more swollen femora in the females of Anteoninae. In Dryininae and Gonatopodinae the fore legs not only have more swollen femora, but also longer coxae and trochanters. The fore legs are therefore longer in females than in males and are longer than the mid and hind legs. This line of evolution is not discernible in Embolemidae and Aphelopinae, where host feeding is not practised and where the chelae are absent. Evolutionary change is also visible in the chelae which progress from a plesiomorphic condition of extreme simplicity and scarce mobility to apomorphic conditions of structural complication and great mobility. The chela of Anteoninae is very simple and slightly movable; the enlarged claw does not have teeth, lamellae or bristles. On the other hand, the chela of Dryininae and Gonatopodinae is more complicated and has teeth, lamellae and bristles. In Gonatopodinae and Dryininae, we also observed more specific evolutionary change, with a progression from enlarged claws with bristles and small teeth to enlarged claws with bristles, lamellae and large teeth.

Abdomen (Figs 8-9)

The abdomen consists of ten segments, but this number may not be visible without careful study or dissection, because the last segments are usually retracted. Each segment is normally made up of a dorsal plate or tergite and a ventral plate or sternite. The spiracular openings lie at the sides of the tergites. The first abdominal segment is modified into the propodeum (Figs 8 B, 9 F) which is functionally part of the thorax. The term 'gaster' or 'metasoma' is used for all the abdominal segments behind the propodeum. The petiole (Fig. 8 C) is the narrowed, anterior, more or less long stalk. In Dryinidae and Embolemidae it is composed of most of the second abdominal segment.

Genitalia (Figs 17-20)

The female genitalia do not seem to be of taxonomic importance. Females of Dryinidae and Embolemidae have an ovipositor, a hollow tube composed of three pairs of valves and two major, basal sclerites. One of the three valves is the second gonocoxa (= second valvifer, sensu Gauld and Bolton, 1988). According to Carpenter (1986) the presence in the second gonocoxa of an articulation which divides the valve into two parts (dorsal and dorsoventral) is an apomorphy of the Chrysidoidea.

The male genitalia seem to be of taxonomic importance mainly in the Dryinidae and only in the subfamilies Aphelopinae, Anteoninae and Gonatopodinae. The male genitalia (Fig. 17) consist of two parameres (= gonoforceps, sensu Olmi, 1984) that are surrounded proximally by a sclerotized basal ring. Ventrobasally they are protected by a small 9th sternite. Projecting internally from the para-

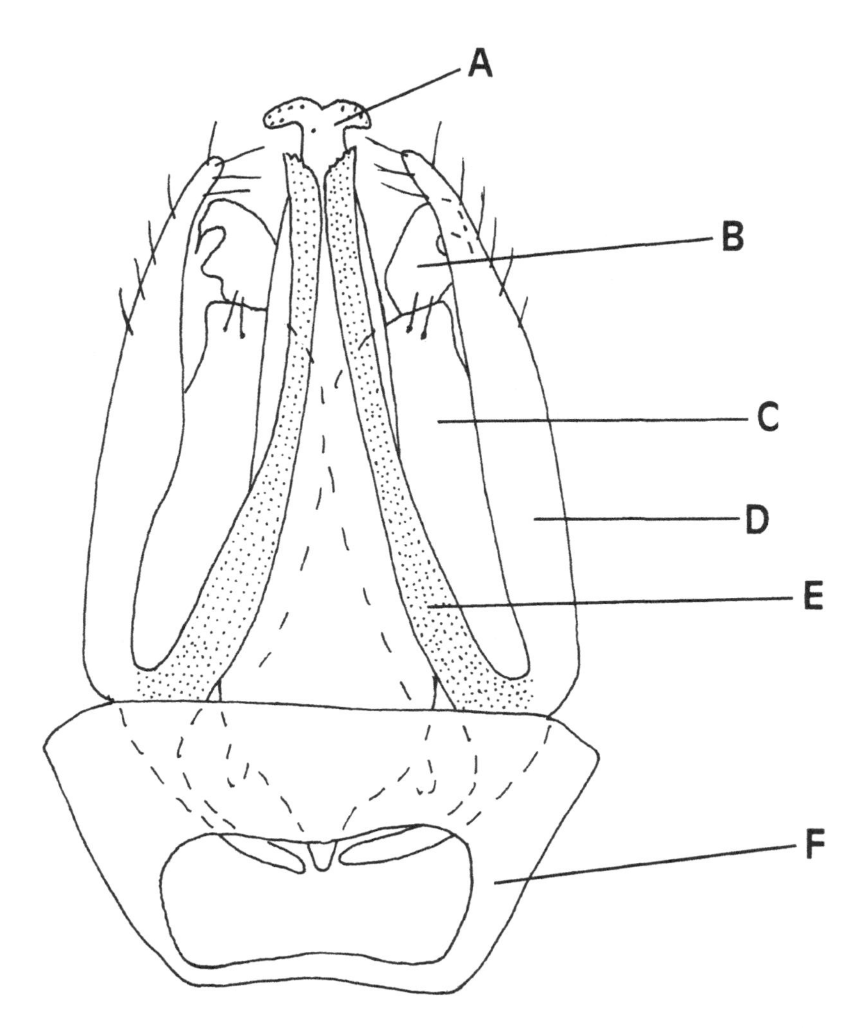

Fig. 17. Male genitalia of Dryinidae: A: aedeagus; B: distivolsella; C: basivolsella; D: paramere; E: dorsal process of parameres; F: basal ring.

meres are the volsellae. Each volsella consists of a proximal large piece called basivolsella and a distal smaller piece called distivolsella. Positioned centrally is the bilobate intromittent organ, the aedeagus (= penis, sensu Olmi, 1984). In the males of Dryinidae Anteoninae and Embolemidae the parameres may have more or less large dorsal membranous lobes proximally (Figs 18 A, 19 A). In the males of the Dryinidae Gonatopodinae the parameres have dorsal processes (Fig. 20 A) which are of great taxonomic importance.

Bionomics: Dryinidae

The Dryinidae are usually both predators and parasitoids of Homoptera Auchenorrhyncha of the Cicadomorpha and Fulgoromorpha (sensu Ossiannilsson, 1978). However, the Cicadidae and Cercopidae are not known as hosts of dryinids. Only the Aphelopinae are not predators.

Dryinids are endoparasitoids in the early instars and ectoparasitoids in the last instars. Only the species of the Nearctic and Neotropical genus *Crovettia* are fully endophagous.

Dryinids live in all terrestrial habitats from sealevel to the high mountains, wherever their hosts are to be found. In Fennoscandia we know of dryi-

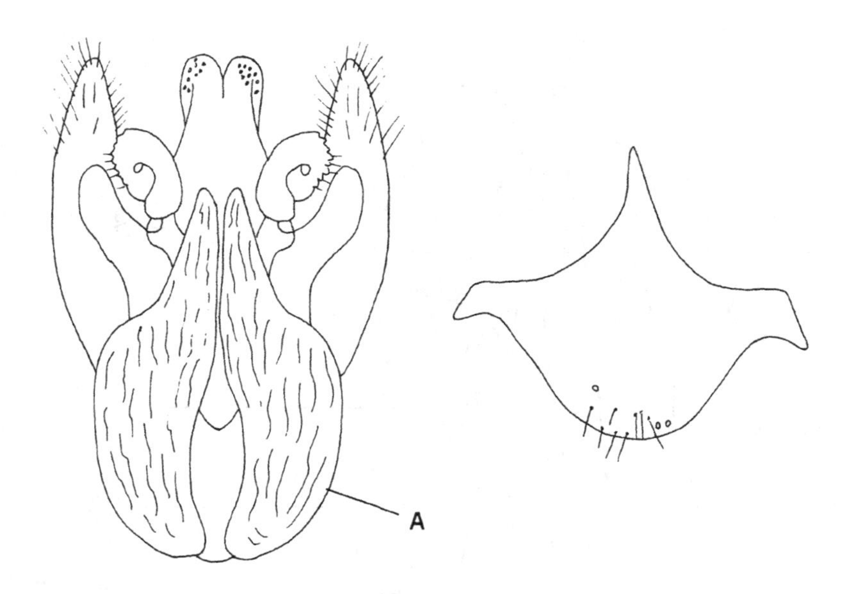

Fig. 18. Male of *Embolemus ruddii* Westwood: genitalia on left; 9th abdominal sternite on right; A: dorsal membranous process.

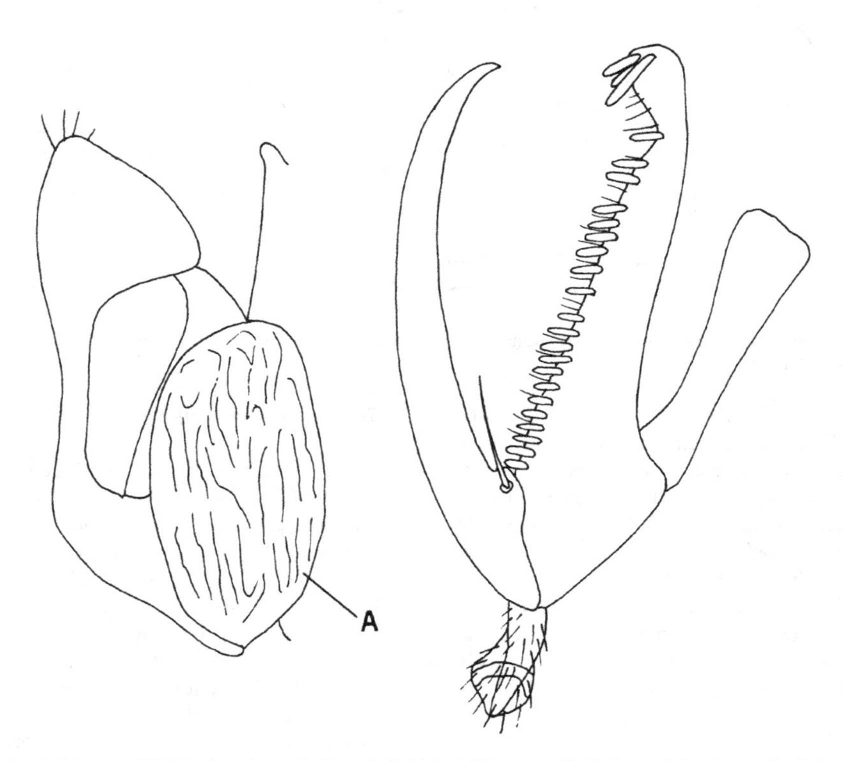

Fig. 19. *Anteon fulviventre* (Haliday): male genitalia on left (right half removed); chela on right; A: proximal dorsal membranous process.

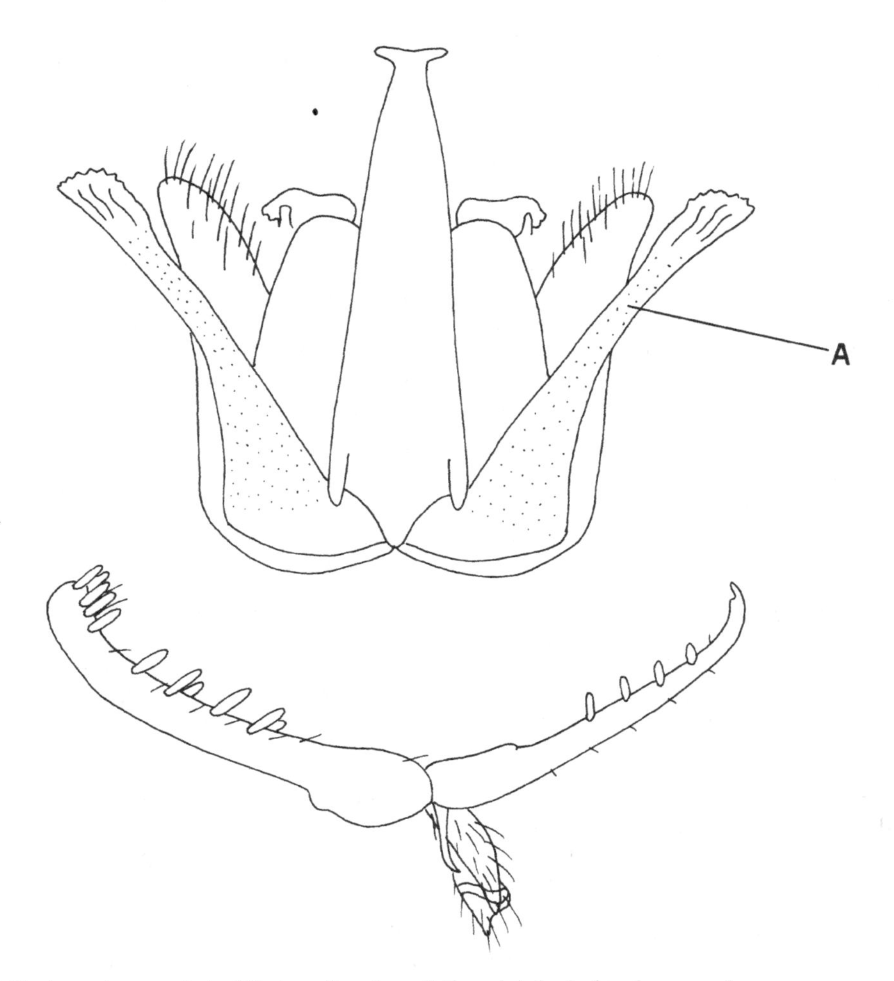

Fig. 20. *Haplogonatopus oratorius* (Westwood): male genitalia and chela; A: dorsal process of parameres.

nids from most northern regions, such as the record of *Lonchodryinus ruficornis* (Dalman) from 71 ° N at Gunnarnes (Rolvsøy, Norway): this is most northerly known occurrence of Dryinidae anywhere in the world.

Dryinid reproduction may be bisexual or parthenogenetic; parthenogenesis may be thelytokous or arrhenotokous.

Males do not feed, or feed only on sugar solutions, mainly the honey dew of their hosts; they live a few days, and commonly die just after mating. Females feed on sugar solutions (mainly honey dew), but they feed mainly on their hosts. In fact, the female grasps the host with its chelae and consumes haemolymph and tissues. The prey may die or survive, according to the severity of its wounds. Host-feeding is absent only in the non-chelate Aphelopinae; according to Jervis, Kidd and Sahragard (1987), this suggests that host-feeding is associated with the possession of chelae. In more specialised subfamilies (Dryininae, Gonatopodinae), host-feeding commonly kills the prey and is called 'destructive' (Jervis and Kidd, 1986). In less specialised subfamilies (Anteoninae) the prey commonly survives ('non-destructive' host-feeding). If it is followed (or not followed) by oviposition, host-feeding may then be 'concurrent' (or 'non-concurrent'). Host-feeding is very important for the control of Auchenorrhyncha, because it kills a large number of hosts, at least equivalent to the number killed through parasitization.

The male dryinid is very lethargic: either it does not move, or it moves only to search for the opposite sex. On the other hand, the female is constantly and rapidly on the move, searching for hosts. In the Aphelopinae the females do not have chelae and do not feed on their hosts; they grasp the body of the host between the two fore legs, with or without the help of their mandibles. In other subfamilies the females capture their hosts with their chelae.

Though the Dryinidae are classified in the Aculeata, their ovipositor is used for egg laying. Oviposition is accompanied by the emission of a paralysing substance. The paralysis does not last long, and the hopper soon recovers and lives normally until the dryinid larva kills it by consuming the haemolymph. Eggs are deposited at different places on the host's body, according to the subfamilies.

In the Aphelopinae there are two types of development, according to the genus. In the Nearctic and Neotropical genus *Crovettia* Olmi, development is completely internal and polyembryonic. In the genus *Aphelopus* Dalman, it is monoembryonic and partly internal and partly external.

Species of the genus *Aphelopus* parasitize only nymphs or adults of the Cicadellidae Typhlocybinae (sensu Ossiannilsson, 1981). The eggs are laid inside the host body. The dryinid female, in fact, stings an intersegmental membrane and lays an egg in the haemocoel. This small alecithal egg divides and forms an embryo and a membrane called a 'trophamnion', which surrounds the embryo. The egg chorion dissolves and the 1st instar larva is exposed to the host body cavity. It is constantly separated from the haemolymph by its trophamnion, through which respiratory and food exchanges occur. Respiration is cutaneous as the tracheae are filled with liquid (Buyckx, 1948). The 2nd instar larva at first remains motionless near the wound inflicted by the adult ovipositor. Subsequently, however, and under pressure from the host's alimentary canal, it is compressed against the host integument. As a result it is expelled from the host through the ovipositor wound. Its body becomes visible on the outside, where, however, it remains concealed between two overlapping gaster sclerites. The entire posterior part of the larval body is external while the head is internal, inside the haemocoel. As the two overlapping sclerites move, the larva runs the risk of injury. To avoid this, a cuticular ring is formed around its head. During the 2nd moult this ring separates together with the old cuticle and fits into the integument hole. Furthermore, the posterior extremity of the larval body remains attached to the host's integument by a viscous substance secreted by the larva. The old cuticles that separate during the 2nd and following moults will all adhere to the host's integument with the same viscous substance. The subsequent cast larval skins that separate during each moult form an external cyst or sac, which projects from the host's body and is called a 'thylacium'. At first it is small, but it grows after each moult. Its colour may be yellow, brown, green or black, according to the host. This cyst, elongate and subrectangular, is very different from the spherical thylacium of other dryinid subfamilies. When the posterior part of the larval body is pushed out of the host, the epidermis of the host starts a cicatricial reaction around the parasitoid larva. This cicatricial tissue is anomalous because it is composed of very large cells. They penetrate between the trophamnion and the larval cuticle and form a layer completely surrounding the dryinid larva. The trophamnion surrounding the posterior part of the larval body then dissolves and in the thylacium the larva is surrounded only by the cicatricial tissue. The trophamnion only remains around the head, in the haemocoel, where it continues its food and respiration exchanges. The cicatricial tissue is interrupted in front of the larval face: through this hole the larva sucks the food transmitted by the trophamnion. Respiration continues to be cutaneous because the parasitoid tracheae are filled with liquid. During the 4th instar the larval cuticle thickens. During the 4th moult this thick larval skin is not cast, but forms a sheath around the 5th instar larva. The 5th instar larva is hymenopteriform (sensu Gauld and Bolton, 1988): it is an eucephalous larva with developed mouth-parts and tracheal respiration. On the other hand, the larva of the earlier instars is sacciform (sensu Gauld and Bolton, 1988), a featureless ovoid larva with rudimentary mouth-parts and cutaneous respiration. The development of the Aphelopine larvae is therefore hypermetabolous. These marked morphological and anatomical transformations occur in the early 5th instar larva, sheltered within the 4th instar cuticle. After the transformations have been completed, the 4th instar larval skin ruptures and the mature larva emerges. The host's behaviour now changes; it becomes sluggish and clings to the food-plant by its rostrum. The mature dryinid larva eats out all the body-contents of the host, which dies. The larva then splits open the thylacium, crawls out and pupates in the soil in a silk cocoon (Buyckx, 1948; Jervis, 1980a, b; Keilin and Thompson, 1915). There are different opinions as to the number of larval instars: some authors refer to 4 instars, others to 5. This problem really derives from the chorion dissolution. There is no process of eclosion from the egg: after the chorion dissolution the organism, surrounded by its trophamnion, may be considered either as an embryo (and then the number of instars is 4) or as a 1st instar larva (and then the number of instars is 5).

Species of the subfamily Gonatopodinae parasitize Fulgoromorpha and Cicadomorpha (sensu Ossiannilsson, 1978). Cicadidae, Cercopidae, Membracidae and Cicadellidae Typhlocybinae do not appear to be parasitized by Gonatopodinae. The development of the Gonatopodinae is very different from that of the Aphelopinae. The small alecithal egg may be glued on to an intersegmental membrane of the host's thorax or abdomen, either on the external (Fenton, 1918a) or internal (Giri and Freytag, 1989) surface. Nymphs or adults of the hosts may be parasitized. The dryinid female stings an intersegmental membrane of the host, which is temporarily paralysed. If the egg is external, the 1st instar larva penetrates into the host just after the eclosion, through the ovipositor wound. The 1st instar larva (deriving from an internal or external egg) penetrates into the ovipositor wound and remains with the posterior part of its body projecting outside the body of the host between two overlapping sclerites. In gonatopodine larvae, the trophamnion and the cicatricial tissue that are characteristic of *Aphelopus* larvae are absent. Respiration is tracheal because the spiracles are open and the tracheae are not filled with liquid, as was the case in *Aphelopus* larvae (Ponomarenko, 1975). The posterior part of the larva is external; it therefore takes in atmospheric air. The mouthparts are very rudimentary and incapable of feeding directly on the host haemolymph. Two oval vesicles are visible in front of the larval face; their function is not clear, but they are probably equivalent to the *Aphelopus* trophamnion and are mediators of food (Ponomarenko, 1975). During the 1st moult the larval cuticle is cast outside the host body and remains around the gonatopodine larva to form the first piece of the thylacium. During the following moults, other cast larval skins are added and the thylacium becomes spherical (Plate 1). Its colour may be black, brown, yellow or green, according to the host species. The penultimate instar larva, before its moult, devours all the body-contents of the host, which dies (Hernandez, 1984; Giri and Freytag, 1989). Just before activity ceases, the host clings to the food-plant by its rostrum. In other gonatopodine species the last instar devours all the host's body-contents (Fenton, 1918a). In Gonatopodinae too the last instar larva is hymenopteriform, with well-developed mouth-parts, whereas the previous instars are sacciform. Development is thus hypermetabolous. The mature dryinid larva splits open the thylacium, crawls out and pupates in a silk cocoon on leaves or stems of the food-plant of its host; occasionally it pupates in the soil. The number of larval instars may be 4 (Giri and Freytag, 1989; Heikinheimo, 1957; Raatikainen, 1961) or 5 (Fenton, 1918b; Kitamura, 1985; Chandra, 1980b).

Species of the subfamily Dryininae parasitize Fulgoromorpha (sensu Ossiannilsson, 1978). Their development is less well known than that of the Aphelopinae and Gonatopodinae. It seems, however, that their development is similar to that of the Gonatopodinae (Haupt, 1932; Gonzalo Abril, 1988). There appear to be 4 larval instars. The cocoon (Plate 2) is spun on leaves or stems.

Species of the subfamily Anteoninae parasitize Cicadomorpha (sensu Ossiannilsson, 1978). However, Cicadidae, Cercopidae, Membracidae and Cicadellidae Typhlocybinae do not seem to be parasitized by Anteoninae. The development of the Anteoninae is little known; it seems to be intermediate between that of *Aphelopus* and the Gonatopodinae and Dryininae (Ponomarenko, 1975). The larva is surrounded by a similar but much thinner trophamnion tissue. Two conical epithelial processes are visible on the side of the mouth; they seem to be equivalent to the oval vesicles of the larvae of Gonatopodinae and Dryininae. The mature larva pupates (Plates 3-4) in a silk cocoon in the soil or on leaves or stems of its host's food-plant. There appear to be 5 larval instars (Fenton, 1918b).

The development of dryinids within their hosts may induce morphological, anatomical and physiological changes. The host's reactions are different according to the dryinid subfamily and to the development stage at the time of the parasitization. If a hopper is parasitized in the adult stage, it does not exhibit fundamental changes because its organs are fully developed; at most its reproductive capacity may be reduced. On the other hand, many changes may take place if the hopper is parasitized in a nymphal stage. It depends on whether the host's metamorphosis is stopped or not. The situation varies from subfamily to subfamily.

In the Aphelopinae the hosts are always nymphs of the Cicadellidae Typhlocybinae. The aphelopine larva does not stop the host metamorphosis. The adult hopper thus exhibits pronounced morphological, anatomical and physiological changes; externally, depigmentation and colour changes may be visible; internally, degeneration of the ovarioles and of the female and male genital ducts. Male adults may partly assume the appearance of females. Male stridulatory organs may be lost. The rate of development may be retarded.

The external genital appendages may be considerably modified in both sexes. The adults thus exhibit 'parasitic castration'.

The Gonatopodinae may parasitize adult or nymphal stages. If the host is a nymph, its metamorphosis may or may not be stopped. If it is stopped, the nymph may not reach the adult stage; if, on the other hand, metamorphosis is not stopped, the nymph becomes adult and in this case it shows morphological, anatomical and physiological changes equivalent to those caused by the Aphelopinae (Giri and Freytag, 1989; Kitamura, 1983, 1988; Lindberg, 1950; Raatikainen, 1967). If the hopper is parasitized in the adult stage, there are no changes.

The Dryininae parasitize their hosts in the adult and the nymphal stages. In the few species studied, metamorphosis was stopped, so that no change was exhibited in the adults (Pillault, 1951; Subba Rao, 1957; and personal observations). The effect of parasitization on the hosts of the Anteoninae has not been reported. According to personal observations by the author, it seems that metamorphosis is stopped. If the hopper is parasitized in the adult stage, there are no changes.

According to Pillault (1951), the metamorphosis is not stopped by a substance produced by the dryinid female and introduced during oviposition, but by the dryinid larva. If the larva lives, then the metamorphosis stops; if the larva dies, then the metamorphosis process recommences. The phenomenon also depends on the individual host reaction to the activity of the dryinid larva. The metamorphosis inhibition may also occur in superparasitism (when there is more than one dryinid larva in the same host) and in multiparasitism (when the dryinid larva is accompanied by the larvae of other parasitoids such as Pipunculidae and Strepsiptera). In fact, the last enclosed parasitoid larva stops the metamorphosis of the other larvae living in the same host (Ponomarenko, 1971, 1975). Occasionally, however, the metamorphosis of the other dryinid larvae is not stopped and more than one mature larva then emerges from the host (not more than two, however). In such cases, the dryinid adults are small and display anomalous colours.

In temperate regions dryinids commonly have 1-3 generations each year. Hibernation takes place in the stages of prepupa, pupa or mature larva in the cocoon. A few species overwinter as first instar larvae within overwintering hosts. Cases of aestivation are also known (Jervis, 1980a).

The dispersal of dryinids depends little on their active movements, for they fly little even if they are fully winged. Their dispersal is dependent on the activity of their parasitized hosts. Macropterous Auchenorrhyncha are known for their long migrations. According to Kitamura and Nishikata (1987), the flight performance of macropterous hosts parasitized by dryinids is not reduced by parasitization.

Many dryinids, and mainly the apterous species, are ant-like in their general appearance. This resemblance is useful when dryinids approach their hosts, because ants, which frequently feed on the honey dew produced by the Auchenorrhyncha, do not attack them. The Australian *Anteon myrmecophilum* (Perkins), however, appears to be definitely myrmecophilous (Perkins, 1905).

Bionomics: Embolemidae

The Embolemidae appear to parasitize nymphs of Achilidae (Homoptera Auchenorrhyncha Fulgoromorpha). However, their biology is insufficiently known; only that of the Nearctic species *Ampulicomorpha confusa* Ashmead has been described (Bridwell, 1958; Wharton, 1989). According to these authors, the hosts of this species are nymphs of *Epiptera floridae* (Walker) (Achilidae), which feed on bracket fungus growing on rotten trees. According to Wharton (1989), the parasitized host nymphs have an external sac similar to that of Dryinidae and composed of embolemid larval exuviae. The embolemid larva pupates in a silk cocoon under tree bark.

The biology of the only species of Embolemidae in Fennoscandia and Denmark, *Embolemus ruddii* Westwood, is not known.

Parasites

We do not know of any parasites of Embolemidae.

The Dryinidae, on the contrary, may be parasitized by hyperparasitoid Hymenoptera of the families Diapriidae, Encyrtidae, Ceraphronidae, Pteromalidae and Chalcididae. In their turn, the hyperparasitoid Encyrtidae may be parasitized by Eulophidae (Perkins, 1906).

The Diapriidae may parasitize Anteoninae and Aphelopinae (Jervis, 1979; Masner, 1976; Waloff, 1975); Ceraphronidae parasitize Gonatopodinae (Chandra, 1980b; Pagden, 1934; Swezey, 1908); Encyrtidae parasitize Gonatopodinae and Dryininae (Olmi, 1984); Pteromalidae parasitize Gonatopodinae (Chandra, 1978; Pagden, 1934); Chalcididae parasitize Dryininae (Delvare and Bouček, 1992).

The activity of the hyperparasitoids may greatly reduce the efficiency of the dryinids.

Economic importance of Dryinidae and Embolemidae

The Dryinidae may parasitize and prey on species of Auchenorrhyncha that are pests of plants; they may therefore be beneficial insects.

Dryinids have been used in biological control programmes in three instances. In 1906 and 1907 *Pseudogonatopus* (= *Gonatopus*) *hospes* Perkins and *Haplogonatopus vitiensis* Perkins were introduced into Hawaii from China and Fiji respectively, for the control of the Sugarcane Planthopper *Perkinsiella saccharicida* Kirkaldy (Swezey, 1928). This introduction failed, mainly because certain hyperparasitoids of Dryinidae (which included the Encyrtid *Saronotum americanum* Perkins) had been accidentally established during the preceding few years. They greatly reduced the efficiency of the introduced dryinids (Williams, 1931). In 1935, *Aphelopus albopictus* Ashmead (= *typhlocybae* Muesebeck, sensu Olmi, 1984) was introduced into New Zealand from the U.S.A. for the control of the Yellow Apple Leafhopper *Edwardsiana crataegi* (Douglas) (= *Typhlocyba froggatti* Baker) (Dumbleton, 1937). At present, further programmes to use dryinids for biological control are under consideration in Nigeria and the U.S.A.

The predatory and parasitic efficiency of dryinids has been studied in many species; predation is usually as important as parasitization for the control of Auchenorrhyncha populations. For instance, Hernandez (1984a, b) found in South America that a female of the Gonatopodine *Haplogonatopus hernandezae* Olmi, an enemy of the Rice Planthopper *Tagosodes orizicolus* (Muir), may kill an average of 47 hoppers by host-feeding and 53 by parasitization during its lifetime (an adult female lives 5-17 days). Commonly, however, the mortality of the Auchenorrhyncha through dryinid parasitization may vary according to locality and season. In some countries, a parasitization rate of up to 78 % of hoppers has been reported (Le Quesne, 1972).

The economic importance of the Embolemidae is not known.

Fossils

Many fossils of the Dryinidae and Embolemidae are known from Cretaceous amber (Taimyr in Siberia: 78-115 m.y.b.p.; Medicine Hat in Canada: 70-75 m.y.b.p.) and Tertiary amber (Dominican amber: 25-40 m.y.b.p.; Baltic amber: 40 m.y.b.p.) (data from Poinar, 1993; see also Olmi, 1984, 1987b).

The following species are known from Baltic amber: Embolemidae: *Embolemus breviscapus* Brues 1933; *Ampulicimorpha succinalis* Brues 1933; Dryinidae: *Thaumatodryinus filicornis* (Brues 1923); *Thaumatodryinus deletus* Brues 1933; *Thaumatodryinus gracilis* Brues 1933; *Lestodryinus mortuorum* Brues 1933; *Lestodryinus vetus* Brues 1933; *Neodryinus somniatus* Brues 1933; *Deinodryinus areolatus* (Ponomarenko 1975); *Dryinus bruesi* (Olmi 1984); *Dryinus balticus* (Olmi 1984).

Brues' species are not recognizable because the type specimens (the only known specimens) appear to have been lost. The original descriptions are not reliable.

Evolution and affinities

The evolution of the Dryinidae was determined by two main pressures: the necessity of capturing a host, and the host's reaction. It must therefore be obvious that evolution has affected mainly the female sex, the only one to make any contact with the host. This also explains why females and males are so different in the dryinids. In the female sex, where the selection pressure was more acute, evolution followed paths that were completely different from those in the males, which were not affected by the same selection pressure. Male selection pressure, on the contrary, was apparently very modest because males do not live very long and pass the few days of their life searching (often unsuccessfully) for a female with whom to mate. Females, on the other hand, may reproduce by parthenogenesis, and mating is not necessary. The modest male selection pressure explains their morphological uniformity and the difficulty in finding distinguishing taxonomic characteristics. It also explains why many genera, based on females and apparently valid because of the obvious differences among females (for instance, among the Gonatopodinae: *Dicondylus* Haliday, *Donisthorpina* Richards, *Pseudogonatopus* Perkins, *Agonatopoides* Perkins, *Apterodryinus* Perkins, *Tetrodontochelys* Richards, etc; among the Dryininae: *Mesodryinus* Kieffer, *Tridryinus* Kieffer, *Perodryinus* Perkins, etc.), were recently considered to be synonyms because of the impossibility of finding distinguishing generic features in the male sex (Olmi, 1993).

Where selection pressure did not affect the females very greatly, males are very similar to females. This is the case in the genus *Aphelopus* Dalman, where the only difference between the two sexes (apart from a few cases of sexual dichroism, as in *Aphelopus melaleucus* (Dalman) and *querceus* Olmi) lies in the antennae: distally thickened (clavate) in females, not distally thickened (filiform) in males. It is clear that the difference in antennal thickening was caused by the necessity of creating space in the antennal segments for the sensilla which facilitate host recognition. In fact, *Aphelopus* species only parasitize Typhlocybinae (Cicadellidae), and have to distinguish them from other Auchenorrhyncha. Apart from this difference in antennal shape, there are no other differences between males and females of *Aphelopus*. This makes it easier to recognize the two sexes, but it also shows that female selection pressure, caused by the necessity of capturing a host, was not very strong. The reason for this probably lies in the behaviour of their hosts, the nymphs of Typhlocybinae (Cicadellidae). Adult hosts are not actually parasitized: if 'thylacia' are found on adult bodies, this follows the nymphal metamorphosis which is not stopped by the parasitoid. The parasitized leafhopper may therefore become adult and does not run the risk of losing its 'thylacium' when moulting because it is firmly attached to the host's body. Nymphs of the Typhlocybinae jump less than do other Auchenorrhyncha: rather, they move laterally when danger threatens. When *Aphelopus* tries to capture them, it encounters few difficulties. Females have not needed to modify part of their fore tarsi into chelae: they approach the host very quickly and grasp it with their front and middle pairs of legs, without needing chelae because the host reaction is not so lively as to need more robust capture systems. Moreover, *Aphelopus* species do not practise host-feeding (typical of all other Dryinidae) and do not require adaptations for host-capture and feeding. They therefore do not need chelae.

The chelae appear in the Anteoninae because the hosts have more lively reactions when danger threatens. More robust devices for capture are needed to prevent the host from freeing itself and escaping after capture. However, evolutionary progress has been gradual. The Anteoninae have two types of chelae.

Firstly, there are the small mobile chelae of *Anteon jurineanum* Latreille, *brachycerum* (Dalman), *arcuatum* Kieffer and *flavicorne* (Dalman). These chelae are slightly mobile, because the fifth fore tarsal segment, one of the two arms of the chelae, has hardly been modified and has few possibilities of movement. The only mobile arm is, therefore, the enlarged claw, which is opposable on the fifth fore tarsal segment. This type of chela ensures that a more secure capture can be made than in *Aphelopus* species though it is less so than with other types of more evolved chelae. The weakness of the capture is also due to the morphological characteristics of the two arms of the chela: the inner margin of the enlarged claw is completely smooth, with no hairs, bristles or lamellae; the inner margin of the fifth fore tarsal segment has only a few bristles (only in *Anteon jurineanum* Latreille are there a few lamellae). In fact, capture is less secure when bristles are present rather than the lamellae found in more evolved chelae. In this case too, the weakness of the chelae is due to the low selection pressure imposed by their hosts. The four species listed above are parasitoids of nymphs of Idiocerinae and Macropsinae (Cicadellidae). Their metamorphosis has ceased and they do not reach the adult stage. When danger threatens, nymphs of Idiocerinae and Macropsinae have livelier reactions than do nymphs of the Typhlocybinae; they jump more readily than do nymphs of Typhlocybinae, but less readily than do most of the other Auchenorrhyncha. The selection pressure of the need to capture a host affected these 4 *Anteon* species more than the *Aphelopus* species, but less than in other dryinids. This explains the low mobility and low evolutionary level of their chela. When the reaction of the hosts was livelier (as in the Deltocephalinae and Iassinae, among the hosts of Fennoscandian and Danish species), the selection pressure gave rise to a more mobile type of chela. This took place in the other Anteoninae, *Lonchodryinus ruficornis* (Dalman) and *Anteon ephippiger* (Dalman), *tripartitum* Kieffer, *pubicorne* (Dalman), *gaullei* Kieffer, *fulviventre* (Haliday) and *exiguum* (Haupt). Their chela is more mobile because the fifth fore tarsal segment has a long prolongation, opposable to the enlarged claw. The inner margin of the fifth fore tarsal segment is provided with many lamellae which ensure a secure capture, even if the inner margin of the enlarged claw is still completely smooth. With this type of chela, *Lonchodryinus* species and these *Anteon* species can capture nymphs and adults of leafhoppers jumping rapidly to escape.

All females of Anteoninae have more robust and longer fore legs than the middle and hind legs, whereas all the legs are similar in *Aphelopus*. The clubs of the femora in the Anteoninae are probably very swollen because they contain the muscles which move the arms of the chela. Evolution has therefore begun to involve not only the chelae but also the fore legs.

In the Dryininae, evolution has taken a further step forwards: in *Dryinus niger* Kieffer the chelae are as mobile as in the more evolved Anteoninae, but host-capture is much more secure. The inner margin of the enlarged claw is provided with numerous lamellae which, together with the numerous lamellae on the fifth fore tarsal segment, help in gripping the prey, an adult Cixiidae. The host is a very robust insect, which jumps rapidly and powerfully. The dryinid thus has to be equally robust and powerful to capture it. And in fact the female of *Dryinus niger* Kieffer has not only robust chelae but also long fore legs and swollen clubs on fore femora. In addition, the pronotum has begun to show a process of disarticulation, not discernible in the Anteoninae: it is becoming elongated and more mobile. There is a pronotal transverse impression which divides the pronotum into two regions, an anterior one and a posterior one. This process is not known in the Anteoninae, where the pronotum is a short, stocky and hardly mobile segment. In *Aphelopus* species, the pronotum is very short, almost invisible in dorsal view, and similar in the two sexes. In the Anteoninae and Dryininae, on the contrary, the pronotum is different in the two sexes, being short and invisible in dorsal view in the males, as in the Aphelopinae. Different selection pressures have thus brought about a differentiation in the pronotum in the two sexes in the Anteoninae and Dryininae.

Finally, in the Gonatopodinae the evolutionary process in the females reaches its climax. Females have not only very mobile and long chelae but also very long fore legs, in which the femora have swollen clubs and the trochanters and coxae are long and slender. Fore legs so long, mobile and slender are, of course, very effective for capturing fleeing prey such as nymphs and adults of Delphacidae and Cicadellidae Deltocephalinae and Aphrodinae. The apterous females of the Gonatopodinae also have a fully disarticulated prothorax, composed of a long and slender pronotum and of propleura that have shifted from a sternal position to a dorsal one, between the head and the pronotum. These modifications increase the length of the first thoracic segment. Furthermore, the entire mesosoma is disarticulated since a narrowing of the scutum appears behind the prothorax. This narrowing increases the slenderness and agility of the body. The dryinid has thus become a perfect machine for capturing prey. Male Gonatopodinae, however, have continued to be completely uniform and to have morphological characteristics similar to those of males in the other subfamilies.

However, different evolutionary stages are evident in females of the Gonatopodinae. The prothorax is less disarticulated in females of *Haplogonatopus oratorius* (Westwood) and *Gonatopus helleni* (Raatikainen), *bicolor* (Haliday) and *pedestris* Dalman. The evolutionary level is lower in the female of the last species, in which the enlarged claw is not provided with lamellae but has only a small tooth and a few bristles. Its fifth fore tarsal segment has few bristles and a few small lamellae. The evolutionary level is higher in the females of the two first species because their enlarged claw has a large tooth and both arms of the chela have many long lamellae.

Among females of the other species of the genus *Gonatopus* Ljungh the evolutionary level is lower in *Gonatopus clavipes* (Thunberg), *lunatus* Klug, *spectrum* (Van Vollenhoven), *striatus* Kieffer, *distinguendus* Kieffer and *formicarius* Ljungh, because of the presence in the enlarged claw of a small tooth and a few bristles. The evolutionary level of females of *Gonatopus distinctus* Kieffer, *dromedarius* (Costa), *formicicolus* (Richards) and *pallidus* (Ceballos) is higher, because their enlarged claws have a large tooth and many lamellae, so that host-capture is much more secure. The last three-named and more highly evolved females are parasitoids of Delphacidae, whereas the less evolved females of *Gonatopus clavipes*, *lunatus*, *spectrum*, *striatus*, *distinguendus* and *formicarius* are parasitoids of Cicadellidae. In the female of *Gonatopus striatus* Kieffer, however, the enlarged claw may have lamellae or bristles. This is an intermediate situation.

The primary evolutionary level of the Gonatopodinae is also similar, because the less evolved females of *Gonatopus pedestris* (Dalman) are parasitoids of Cicadellidae, whilst the more evolved females of *Haplogonatopus oratorius* (Westwood) and *Gonatopus helleni* (Raatikainen) and *bicolor* (Haliday) are parasitoids of Fulgoromorpha.

This discussion clearly shows that the evolutionary process in the Dryinidae has concerned only the female sex, whilst the male sex has remained on the fringes of this process. This gives rise to the following important consequences:

1) we can discuss female affinities, because females show clear and different evolutionary levels;
2) we can discuss male affinities, but with great difficulty since males are very uniform and their differential morphological characteristics are usually very slight;

3) we cannot discuss species affinities, because evolution has followed completely different paths in males and females, and female affinities are completely different from male affinities.

So far as the Embolemidae are concerned, the sister-group of the Dryinidae according to recent theories (see the section 'Classification of the Chrysidoidea'), evolution has not progressed as far as in the Dryinidae. Evolutionary pressure, based on the necessity of capturing a host, cannot have been very great because males and females are rather similar. This resemblance is clearly visible in *Ampulicomorpha* Ashmead (not present in Fennoscandia or Denmark), in which females and males are fully winged. In the second of the two genera of this family, *Embolemus* Westwood, to which the only North European species belongs, *ruddii* Westwood, the resemblance is less obvious, because females are micropterous and males are fully winged. This has given rise to important morphological modifications, mainly in the thorax. However, these differences have been caused by wing reduction and not by the evolutionary pressure imposed by the need to capture a host. Unfortunately, the biology of the Embolemidae is too little known and we do not know how females capture their hosts. The only observations of host capture by Embolemidae are those of Bridwell (1958) and Wharton (1989), on *Ampulicomorpha confusa* Ashmead, a Nearctic species. Bridwell observed females of this wasp as they pursued the nymphs of their host and he saw that the nymphs were 'firmly gripped'. Unfortunately, he did not explain how these nymphs were gripped, whether it was with the fore legs, as I believe, with the help of other legs or the mandibles. Their behaviour is probably like that of *Aphelopus* species. The hosts of the Embolemidae are nymphs of the Achilidae, which have very modest reactions when danger threatens. They jump, but their reflexes are slow. They live in a confined environment, under the bark of decayed trees, and are unable to escape easily because their jumps cannot be very long. Embolemidae females are therefore able to capture their hosts even without chelae, unlike most of the Dryinidae. Wharton's (1989) paper consists mainly of larval descriptions.

A cladogram (Tab. 2) is given to show the possible phylogenetic relationships among females of the Fennoscandian and Danish Dryinidae and Embolemidae.

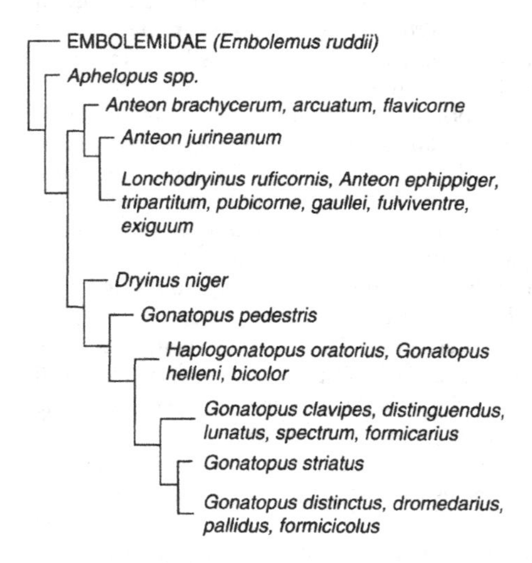

Tab. 2. Phylogenetic relationships among the females of Fennoscandian and Danish Embolemidae and Dryinidae.

Family Embolemidae

Type genus: ***Embolemus*** Westwood, 1833a.

Morphology of both sexes: fully winged (male and female) or micropterous (only female); fore wing with costal, basal, subbasal, first discal and first subdiscal cells fully enclosed by pigmented veins; occasionally first subdiscal cell not completely enclosed (in Fennoscandian and Danish species); first discal and first subdiscal cells enclosed by veins less pigmented than those enclosing costal, basal and subbasal cells; marginal cell open; antennae 10-segmented, articulated on a frontal process; antennal sockets very far from upper margin of clypeus; antennae always more or less geniculated; this character is clearer in females, where antennal segment 1 is very long, much longer than segment 3; in males, antennal segment 1 usually shorter, so that only the female antennae may be considered geniculated; fore tarsus of female not chelate; occipital carina complete; ocelli distinct or absent; tibial spurs 1, 2, 2.

Distribution: worldwide (Olmi, 1994).

Hosts: Achilidae (Homoptera Auchenorrhyncha Fulgoromorpha).

Genera: one genus with one species in Fennoscandia and Denmark.

Genus *Embolemus* Westwood, 1833

Embolemus Westwood, 1833a: 444.
Type-species: *Embolemus ruddii* Westwood, 1833a, by monotypy.
Myrmecomorphus Westwood, 1833b: 496.
Type-species: *Myrmecomorphus rufescens* Westwood, 1833b, by monotypy.
Pedinomma Förster, 1856: 94.
Type-species: *Pedinomma rufescens* (Westwood, 1833), by monotypy.

Female: micropterous; maxillary palps with 3-5 segments; labial palps with 2 segments; third segment of maxillary palps broadened; ocelli absent; pronotum without or with the trace of a median furrow; head pyriform.

Male: fully winged; fore wing usually with costal, basal, subbasal and first discal cells fully enclosed by pigmented veins; first subdiscal cell not fully enclosed by pigmented veins, partly open; pronotum without or with the trace of a median longitudinal furrow; maxillary palps with 3-6 segments; labial palps with 2-3 segments; third segment of maxillary palps broadened; ocelli distinct; head not pyriform.

Distribution: worldwide.

Species: one species in Fennoscandia and Denmark.

1. *Embolemus ruddii* Westwood, 1833

Plates 5-6, Figs 1, 18.

Embolemus ruddii Westwood, 1833a: 445.
Myrmecomorphus rufescens Westwood, 1833b: 496.
Pedinomma rufescens Westwood var. *antennalis* Kieffer in Kieffer & Marshall, 1906: 470.
Embolemus ruddii Westwood var. *rufus* Kieffer in Kieffer & Marshall, 1906: 473.

Female (Plate 5, Fig. 1): micropterous; length 2-5 mm; completely testaceous or testaceous-reddish or brown; antennae geniculate, not thickened distally; according to Hilpert (1989), the proportions of the antennal segments, the ratio between length and breadth of each segment and the ratio between antennae and body length may be very variable; however, usually antennae shorter than body; antennal segment 1 always much longer than segment 3; antennae articulated to strong frontal processes (Fig. 1); antennal sockets very far from upper margin of clypeus; head pyriform, shining, alutaceous, covered with short hairs, without visible sculpture or with fine punctures; occasionally sculpture on sides of head different from sculpture on central region of head; according to Hilpert (1989) this difference in sculpture is only a variation without taxonomic value; occipital carina complete; ocelli absent; only traces of the ocelli are visible; frontal line absent; eyes very small, approximately 0.25 as long as head; region of frons from clypeus to antennal sockets with two longitudinal and median sutures very convergent; these sutures much closer at antennal sockets than at clypeus; these sutures often not completely visible; pronotum alutaceous, shining, without sculpture, covered with short hairs, crossed by a strong transverse impression; anterior collar very short; disc long, with the trace of a median longitudinal furrow; pronotal tubercles reaching tegulae; scutum small, weakly granulated, with the trace of a median furrow, without notauli and parapsidal lines; scutellum very small, alutaceous; metanotum very short, rugose, transverse, fused with propodeum; mesopleura reticulate rugose; meso-metapleural suture complete; metapleura reticulate rugose, fused with propodeum; propodeum reticulate rugose, with spiracles very prominent, with two lateral pointed apophyses on sides of posterior surface; fore wing very reduced, as short as tegulae; hind wings absent; petiole short; maxillary palps with 4 segments; labial palps with 2 segments; third segment of maxillary palps broadened.

Male (Plate 6, Fig. 18): fully winged; length 2.56-3.12 mm; black or completely reddish-brown or reddish-testaceous; antennae and legs brown or testaceous; antennae not geniculated, filiform, not thickened distally; antennae articulated to prominent contiguous processes; antennal sockets very far from upper margin of clypeus; antennal segments in following proportions: 10:4:18:17:16:15:14:14:12:14; the ratio between antennal segments does not seem to vary, except mainly between segments 1 and 3; segment 3 is in fact always much longer than segment 1 (22:11; 16:10; 22:10); head not pyriform, shining, without sculpture, finely haired; occipital carina complete; ocelli distinct; POL = 3; OL = 3; OOL = 6; OPL = 7; TL = 7; frons with the trace of a median furrow from anterior ocellus to frontal antennal process; eyes small, approximately 0.5 as long as head, larger than female eyes; region of frons from clypeus to antennal sockets with two longitudinal and median sutures very convergent; these sutures much closer at antennal sockets than at clypeus, and always completely and distinctly visible; pronotum dull, crossed by a strong transverse impression; anterior collar short; posterior surface (disc) rugose, short, with a strong median longitudinal furrow; pronotal tubercles reaching tegulae; posterior surface of pronotum less than half length of scutum (6:19); pronotum shorter than scutum (9:19); scutum shining, finely punctate, without sculpture among punctures; notauli incomplete, short, reaching approximately 0.25 length of scutum; a very short median mesoscutal sulcus visible near anterior margin of scutum, shorter than notauli; scutellum shining, finely haired and punctate, without sculpture among punctures; metanotum short, transverse, rugose; propodeum dull, reticulate rugose, without transverse or longitudinal keels; mesopleura and metapleura shining, smooth, without sculpture; fore wing hyaline, without dark transverse bands, with 4 cells completely enclosed by pigmented veins (costal, basal, subbasal, first discal); first subdiscal cell partly enclosed by pigmented veins, open; marginal cell open; stigmal vein evenly curved, with distal part longer than proximal part (25:16; 26:16; 29:18); ninth sternite with a short rod (Fig. 18); genitalia with a glabrous dorsal membranous process on parameres (Fig. 18); maxillary palps with 6 segments; labial palps with 3 segments; third segment of maxillary palps broadened.

Distribution: uncommon in Denmark (EJ, SZ, NEZ), Sweden (Sk., Bl., Sm., Öl., Ög., Upl., Vrm., Med.), Norway (Bø, TEi, HOy), East Fennoscandia (Ab, N, Oa, Vib.). -- Widespread, but uncommon, in Europe (Russia, Hungary, Slovakia, Czech Republic, Croatia, Germany, Austria, Switzerland, France, England, Wales, Scotland, Italy). Also in Siberia.

Biology: adults in oak and pine forests from April to October; females also in ant and mole nests or under stones. Probably a parasitoid of *Cixidia* spp. (Achilidae) living under the bark of rotten logs (Olmi, 1994).

Family Dryinidae

Type genus: ***Dryinus*** Latreille, 1804.

Morphology of both sexes: fully winged (male and female) or micropterous (female or rarely male) or apterous (female); fore wing with costal, basal and subbasal cells completely enclosed by pigmented veins (Anteoninae, Dryininae, Gonatopodinae); 1 cell (costal) enclosed in Aphelopinae, 2 (costal and basal) in Conganteoninae; first discal cell completely enclosed only in a few Dryininae (*Thaumatodryinus*); marginal cell open; antennae 10-segmented, not articulated on a frontal process; antennal sockets situated close to upper margin of clypeus; front tarsus in females usually chelate; without chelae only in Aphelopinae females; tibial spurs 1, 1, 2 or 1, 1, 1 or 1, 0, 1 or 1, 0, 2.

Distribution: worldwide.

Hosts: Cicadomorpha and Fulgoromorpha (Homoptera Auchenorrhyncha), except for Cicadidae and Cercopidae.

Subfamilies: 4 subfamilies in Fennoscandia and Denmark (Aphelopinae, Anteoninae, Dryininae, Gonatopodinae).

Genera: 6 genera with 34 species in Fennoscandia and Denmark.

Key to genera of Dryinidae

Females

1 Fore tarsus not chelate (Fig. 21); fully winged (Fig. 21); fore wing with only costal cell completely enclosed by pigmented veins (Fig. 21) .... 1. *Aphelopus* Dalman

– Fore tarsus chelate (Fig. 10); fully winged (Fig. 9) or apterous (Fig. 10); rarely brachypterous (Fig. 28); in fully winged forms fore wing with costal, basal and subbasal cells completely enclosed by pigmented veins (Fig. 31) 2

2 Chela without rudimentary claw (Fig. 34 B); fully winged (Fig. 31); rarely brachypterous (Fig. 28); occipital carina complete 3

– Chela with rudimentary claw (Fig. 42 B); fully winged (Plate 18) or apterous (Fig. 43); occipital carina absent or only visible behind ocellar triangle 4

3 Fore wing with distal part of stigmal vein as long as or longer than proximal part, usually forming a curve between proximal and distal parts (Plate 9); occasionally slightly shorter, but in that case propodeum without a strong transverse keel between dorsal and posterior surfaces (Plate 9) .... 2. *Lonchodryinus* Kieffer

Fore wing with distal part of stigmal vein much shorter than proximal part, usually with an angle between proximal and distal parts (Fig. 31); occasionally slightly shorter or as long as or longer than proximal part, but in that case propodeum with a strong transverse keel between dorsal and posterior surfaces (Plate 13) . . . . 3. *Anteon* Jurine

4 Fully winged (Plate 18); mid leg with a tibial spur (tibial spurs 1, 1, 1 or 1, 1, 2); occipital carina visible behind ocellar triangle . . . . 4. *Dryinus* Latreille

– Apterous (Fig. 43); mid leg without tibial spurs (tibial spurs 1, 0, 1); occipital carina absent 5

5 Palpal formula 2/1 . . . . 5. *Haplogonatopus* Perkins

– Palpal formula different (2/2, 3/2, 4/2, 5/2, 4/3, 5/3, 6/3) . . . . 6. *Gonatopus* Ljungh

Males

1 Fore wing with only costal cell completely enclosed by pigmented veins (Plate 8) . . . . 1. *Aphelopus* Dalman

– Fore wing with costal, basal and subbasal cells completely enclosed by pigmented veins (Fig. 29) 2

2 Mandibles with 1-3 teeth; occipital carina absent or only visible on dorsal side of head, not visible on genae and temples 3

– Mandibles with 4 teeth; occipital carina complete, also visible on lateral sides of head . 5

3 Occipital carina visible on dorsal side of head, touching eyes laterally; parameres without a dorsal process (Fig. 42 A) . . . . 4. *Dryinus* Latreille

– Occipital carina absent or briefly visible on sides of posterior ocelli, never reaching eyes laterally; parameres with a dorsal process (Figs 20, 48 A) 4

4 Palpal formula 2/1 . . . . 5. *Haplogonatopus* Perkins

– Palpal formula different (2/2, 3/2, 4/2, 5/2, 4/3, 5/3, 6/3) . . . . 6. *Gonatopus* Ljungh

5 Fore wing with distal part of stigmal vein as long as or longer than proximal part, usually forming a curve between proximal and distal parts (Fig. 29); occasionally slightly shorter, but in that case propodeum without a strong transverse keel between dorsal and posterior surfaces (Plate 11) . . . . 2. *Lonchodryinus* Kieffer

– Fore wing with distal part of stigmal vein much shorter than proximal part, usually with an angle between proximal and distal parts (Fig. 2); occasionally slightly shorter or as long as or longer than proximal part, but in that case propodeum with a strong transverse keel between dorsal and posterior surfaces (Plate 17) . . . . 3. *Anteon* Jurine

1. Subfamily Aphelopinae

Type genus: ***Aphelopus*** Dalman, 1823.

Morphology of both sexes: fully winged; fore wing with only costal cell completely enclosed by pigmented veins; female fore tarsus not chelate; tibial spurs 1, 1, 2.

Distribution: worldwide.

Hosts: Cicadellidae Typhlocybinae (genus *Aphelopus* Dalman) and Membracidae (genus *Crovettia* Olmi; not present in Fennoscandia and Denmark).

Genera: one genus in Fennoscandia and Denmark.

1. *Genus Aphelopus* Dalman, 1823

Aphelopus Dalman, 1823: 8.

Type-species: *Dryinus atratus* Dalman, 1823, by subsequent designation.

Female: maxillary palps with 5 segments; labial palps with 2 segments; mandibles quadridentate; fore wing with stigmal vein evenly curved; head, scutum, scutellum and metanotum usually completely granulated, not reticulate rugose; antennae thickened distally.

Male: structurally resembling female; antennae not thickened distally.

Distribution: worldwide.

Species: 6 species in Fennoscandia and Denmark.

Notes: female specimens are not always easily identifiable; females of *A. melaleucus* (Dalman),

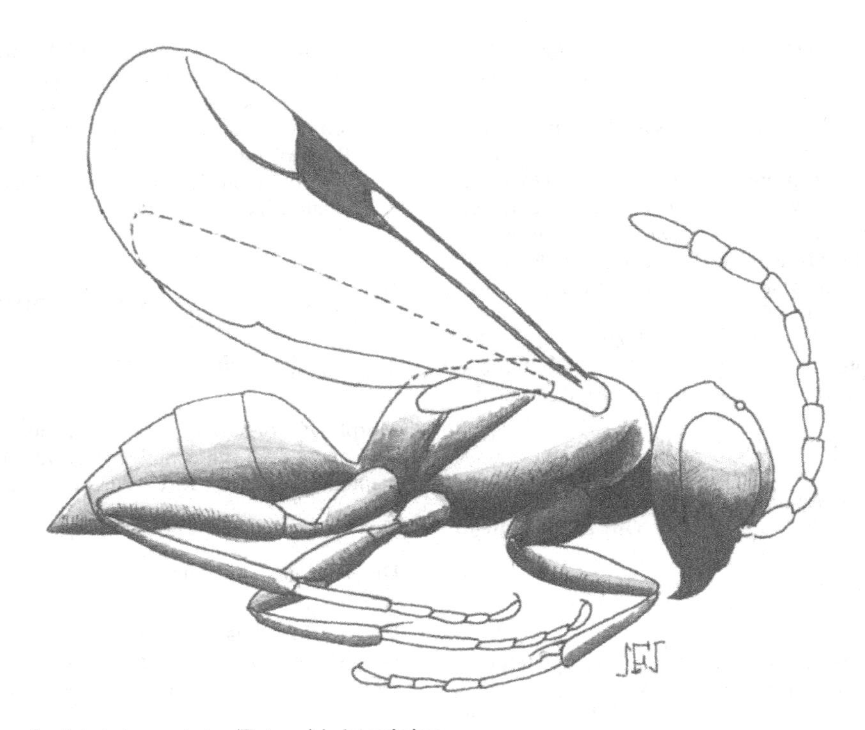

Fig. 21. Female of *Aphelopus atratus* (Dalman) in lateral view.

querceus Olmi and *serratus* Richards can usually be easily recognized; on the other hand, females of *A. camus* Richards, *atratus* (Dalman) and *nigriceps* Kieffer cannot always be easily recognized; in the following key, females of *A. camus* and *nigriceps* are not separable when they have notauli reaching approximately 0.65 length of scutum. The males, on the other hand, can always be easily recognized. Jervis (1977) proposed the following character to distinguish *A. nigriceps* Kieffer: 'epistomal suture and subantennal areas distinctly sculptured'. This is a good character, but more clearly visible from scanning electron micrographs than with an optical microscope.

Key to species of *Aphelopus*

Females

1 Head with mandibles, clypeus and a more or less conspicuous frontal surface white or testaceous (Plate 7 G, H) 2
– Head with at most clypeus and mandibles white or testaceous 3
2 Antennae more thickened distally; frons with a U-shaped white or testaceous mark, enclosing bases of antennae, with arms lying along inner margin of eyes (Plate 7 H); frons rarely with only a few small white or testaceous spots near base of clypeus and mandibles; notauli reaching approximately 0.5 length of scutum .... 2. *melaleucus* (Dalman)
– Antennae distally less thickened; frons with only lower face white or testaceous (Plate 7 G); occasionally frons with a U-shaped white or testaceous mark, enclosing bases of antennae, with arms lying along inner margin of eyes for only a short distance; notauli reaching approximately 0.60-0.75 length of scutum .... 7. *querceus* Olmi
3 Head with only mandibles white or testaceous; clypeus black or uniformly dark 4
– Head with clypeus and mandibles white or testaceous; clypeus occasionally partly dark, never uniformly dark 6
4 Notauli reaching at most 0.6 length of scutum .... 3. *atratus* (Dalman)
– Notauli reaching at least 0.65 length of scutum 5
5 Notauli reaching beyond 0.75 length of scutum, occasionally reaching posterior margin of scutum .... 4. *serratus* Richards
– Notauli reaching approximately 0.65 length

of scutum ....
.... 5. *camus* Richards, 6. *nigriceps* Kieffer

6 Notauli reaching beyond 0.75 length of scutum; occasionally reaching posterior margin of scutum; epistomal suture medially straight (see plate 7 A); width of malar space at its narrowest point equal to, or greater than, median height of clypeus (see plate 7 A) ....
.... 4. *serratus* Richards

– Notauli reaching approximately 0.65 length of scutum; epistomal suture medially curved (Plate 7 D); width of malar space at its narrowest point distinctly less than medial height of clypeus (Plate 7 D) .. 5. *camus* Richards

Males

1 Aedeagus distally tridentate (Fig. 23 A) ....
.... 3. *atratus* (Dalman)

– Aedeagus distally not tridentate (Figs 22, 24-27) 2

2 Distivolsella in the form of a long straight rod (Fig. 24 A); basivolsella long and narrow, pointed distally (Fig. 24 A) ....
.... 4. *serratus* Richards

– Distivolsella differently formed (Figs 22, 25-27); basivolsella broader (Figs 22, 25-27) ...
.... 3

3 Basivolsella with 1 subdistal bristle (Figs 22, 25) 4

– Basivolsella with 2 subdistal bristles (Figs 26, 27) 5

4 Aedeagus with distal apex trumpet-shaped (Fig. 25 A) .... 5. *camus* Richards

– Aedeagus with distal apex not trumpet-shaped (Fig. 22 A)
.... 2. *melaleucus* (Dalman)

5 Basivolsella with an outer basal process (Fig. 27 A) .... 7. *querceus* Olmi

– Basivolsella without an outer basal process (Fig. 26 A) .... 6. *nigriceps* Kieffer

2. *Aphelopus melaleucus* (Dalman, 1818)

Plates 7-8, Fig. 22.

Gonatopus melaleucus Dalman, 1818: 82.
Aphelopus albipes Kieffer in Kieffer & Marshall, 1905: 217.
Aphelopus trisulcatus Kieffer, 1914: 216.

Female: fully winged; length 1.5-2.0 mm; black; mandibles, clypeus, malar space, lower face and a U-shaped mark (enclosing bases of antennae, with arms lying along inner margins of eyes) (Plate 7 H) white or testaceous; frons rarely with only a few small white or testaceous spots near base of clypeus and mandibles, as in males; antennae completely testaceous or brown, with segments 8-10 light; legs testaceous-whitish, with hind coxae, hind clubs of femora and hind tibiae partly brown; occasionally also mid coxae and clubs of mid and fore femora partly brown; antennae thickened distally; antennal segments in following proportions: 4:3.5:4.5:5:5:5:4:4.5:4:7; head dull, smooth, completely granulated; frontal line present or absent; occipital carina complete; POL = 4; OL = 2; OOL = 2.6; OPL = 2; TL = 2; scutum, scutellum and metanotum dull, smooth, granulated; notauli incomplete, reaching approximately 0.5 length of scutum; propodeum reticulate rugose, with 2 longitudinal keels on posterior surface; median area shining, less rugose than lateral regions, smooth, with a few irregular weak keels, without areolae; fore wings hyaline, without dark transverse bands.

Male (Plate 8): fully winged; length 1.75-2.37 mm; black; mandibles, clypeus and a few small spots between antennal sockets and on malar space yellowish or whitish (Plate 7 B); occasionally head completely black, without light spots, with only mandibles testaceous; antennae brown or black, occasionally with segments 9-10 light; legs brown, with tarsi and fore tibiae testaceous; antennae not thickened distally; antennal segments in following proportions: 3.5:4:4.5:5:5.5:6:7:7:7:10; head dull, smooth, granulated; frontal line present; occipital carina complete; POL = 7; OL = 3; OOL = 4; OPL = 2; TL = 2; scutum, scutellum and metanotum dull, smooth, granulated; notauli incomplete, reaching approximately 0.5 length of scutum; occasionally scutum with a median mesoscutal sulcus; propodeum reticulate rugose, with two longitudinal keels on posterior surface; median area shining, less rugose than lateral regions, almost smooth, with a few weak irregular keels; fore wing hyaline, without dark transverse bands; ninth abdominal sternite strongly emarginate posteriorly (Fig. 22 B); basivolsella with 1 subdistal bristle, without an outer process (Fig. 22 A); aedeagus with apex not trumpet-shaped, not tridentate (Fig. 22 A).

Distribution: common in Denmark (SJ, EJ, F, LFM, SZ, NEZ, B), Sweden (Sk., Bl., Sm., Gtl., Ög., Vg., Nrk., Upl., Dlr., Hls., Jmt., Vb.), Norway (Ø, AK, Bø, VE, TEi, AAy, HOy), East Fennoscandia (Al, Ab, N, Ta, Sa, Tb, Ok, Ks, Li, Vib, Kr). Widespread in Europe, also in Siberia, Armenia, Cyprus, Lebanon.

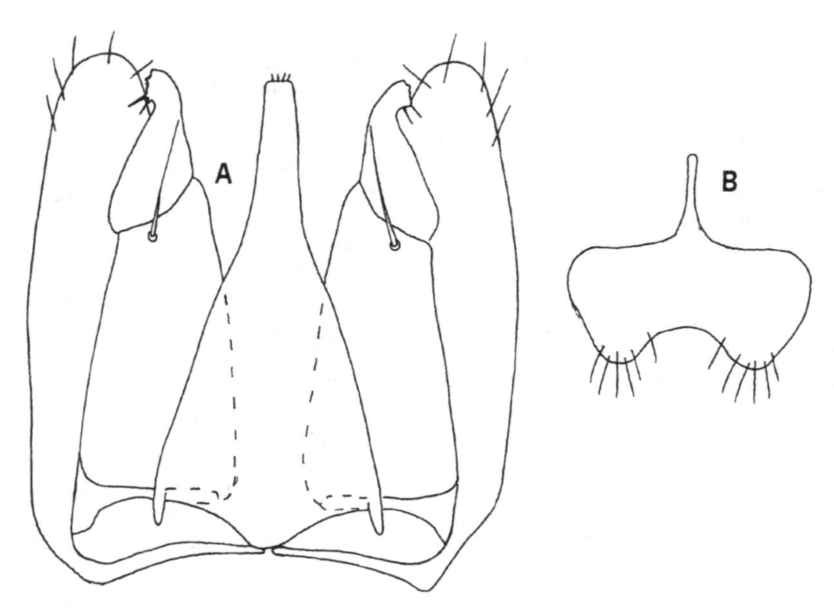

Fig. 22. Male of *Aphelopus melaleucus* (Dalman): genitalia (A) and 9th abdominal sternite (B).

Biology: adults in broadleaf forests from May to September; also in fields and pastures. A parasitoid of numerous species of Typhlocybinae (Cicadellidae). Reared from *Empoasca vitis* (Göthe) and *decipiens* Paoli; *Fagocyba cruenta* (H.-S.), *douglasi* (Edwards) and *carri* (Edwards); *Ossiannilssonola callosa* (Then); *Edwardsiana rosae* (L.), *avellanae* (Edwards), *crataegi* (Douglas), *menzbieri* Zachvatkin, *flavescens* (F.), *plebeja* (Edwards), *geometrica* (Schrank), *bergmani* (Tullgren), *hippocastani* (Edwards), *lethierryi* (Edwards); *Linnavuoriana decempunctata* (Fallén); *Ribautiana ulmi* (L.); *Typhlocyba quercus* (F.); *Aguriahana germari* (Zetterstedt); *Alnetoidia alneti* (Dahlbom); *Zygina flammigera* (Fourcroy). The prepupa overwinters in the soil in a cocoon. The number of generations in Fennoscandia and Denmark is not known, but in England the species is facultatively bivoltine or univoltine (Jervis, 1980b, c). Courtship and mating were described by Jervis (1979b). It may be parasitized by *Ismarus dorsiger* (Curtis) (Hymenoptera Diapriidae) (Jervis 1979a).

3. ***Aphelopus atratus*** (Dalman, 1823)

Plate 7, Figs 21, 23.

Dryinus (Aphelopus) atratus Dalman, 1823:15.
Aphelopus holomelas Richards, 1939: 289.

Female (Fig. 21): fully winged; length 1.5-2.0 mm; black; mandibles testaceous; antennae completely black, occasionally with segment 1 or segments 1-2 testaceous; legs completely testaceous, occasionally with hind femoral clubs partly brown; also hind coxae occasionally brown; antennae thickened distally; antennal segments in following proportions: 5:4.5:4.5:6:6:6:5:5:5:8; head dull, completely granulated; frontal line present; occipital carina complete; POL = 7; OL = 3; OOL = 4; OPL = 4; TL = 3; scutum, scutellum and metanotum dull, granulated; notauli incomplete, reaching approximately 0.5 length of scutum; occasionally median mesoscutal sulcus present; propodeum reticulate rugose, with two longitudinal keels on posterior surface; median area less rugose than lateral regions, not reticulate rugose, with a few weak irregular keels; fore wing hyaline, without dark transverse bands.

Male: fully winged; length 1.50-2.25 mm; black; mandibles testaceous (Plate 7 C); antennae completely black or brown; legs brown, with tarsi and fore tibiae testaceous; antennae not thickened distally; antennal segments in following proportions: 4:4:5:6:6.5:7:7:8:8:12; head dull, granulated; frontal line present or absent; occipital carina complete; POL = 8; OL = 5; OOL = 5; OPL = 3; TL = 3; scutum, scutellum and metanotum dull, granulated; notauli incomplete, reaching approximately 0.5 length of scutum; occasionally median mesoscutal sulcus present; propodeum reticulate rugose, with two longitudinal keels on posterior surface;

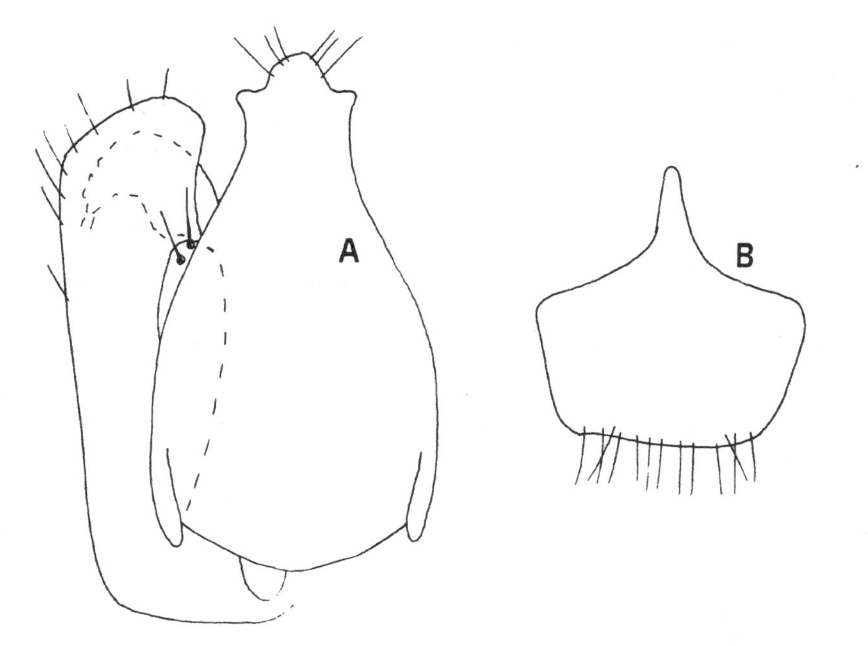

Fig. 23. Male of *Aphelopus atratus* (Dalman): genitalia (right half removed) (A) and 9th abdominal sternite (B).

median area smooth, less rugose than lateral regions, with a few weak irregular keels; fore wing hyaline, without dark transverse bands; ninth abdominal sternite not emarginated posteriorly (Fig. 23 B); distal apex of aedeagus tridentate (Fig. 23 A); basivolsella with 2 (rarely 3) subdistal bristles (Fig. 23 A).

Distribution: common in Denmark (EJ, F, SZ, NEZ), Sweden (Sk., Bl., Sm., Öl., Ög., Vg., Boh., Nrk., Sdm., Upl., Vstm., Hls.), Norway (Ø, AK, Bø, VE, TEi), East Fennoscandia (Al, Ab, N, St, Sa, Sb, Kb, Vib). Widespread in Europe, also in Siberia, Armenia, Cyprus.

Biology: adults in broadleaf forests from May to August; also in pastures and fields. A parasitoid of numerous species of Typhlocybinae (Cicadellidae). Reared from *Alebra albostriella* (Fallén) and *wahlbergi* (Boheman); *Empoasca vitis* (Göthe); *Fagocyba cruenta* (H.-S.); *Edwardsiana rosae* (L.), *crataegi* (Douglas), *hippocastani* (Edwards) and *lethierryi* (Edwards); *Ribautiana ulmi* (L.) and *tenerrima* (H.-S.); *Typhlocyba quercus* (F.) and *bifasciata* Boheman; *Eupteryx aurata* (L.), *urticae* (F.), *cyclops* Matsumura and *stachydearum* (Hardy); *Zygina flammigera* (Fourcroy). The prepupa overwinters in the soil in a cocoon. The number of generations in Fennoscandia and Denmark is not known, but in England the species may be univoltine or bivoltine (Waloff and Jervis, 1987). Information on biology and larval morphology in Buyckx (1948; misid. *Aphelopus indivisus* Kieffer). The species may be parasitized by Encyrtidae (Buyckx, 1948).

4. ***Aphelopus serratus*** Richards, 1939

Plate 7, Fig. 24.

Aphelopus serratus Richards, 1939: 284.

Female: fully winged; length 1.5-2.5 mm; black; mandibles and clypeus testaceous or whitish; occasionally clypeus uniformly black or dark; occasionally head partly reddish-brown; antennae black; legs testaceous, with hind coxae and hind femora partly brown; occasionally mid femora partly brown; antennae thickened distally; antennal segments in following proportions: 6:4:6:6:6:6:5:6:5:8; head dull, completely granulated; frontal line present; occipital carina complete; POL = 7; OL = 4; OOL = 3.5; OPL = 3; TL = 3; epistomal suture straight medially (see plate 7 A); width of malar space at its narrowest point equal to or greater than medial height of clypeus (see plate 7 A); scutum, scutellum and metanotum dull, granulated; notauli incomplete, reaching beyond 0.75 length of scutum; occasionally complete and reaching posterior margin of scutum, separated; median mesoscutal sulcus occasionally present; propodeum reticulate rugose, with two longitudinal keels on posterior surface; median area reticulate rugose; fore wing hyaline, without dark transverse bands.

Male: fully winged; length 1.75-2.12 mm; black; mandibles and clypeus testaceous or whitish (Plate 7 A); occasionally clypeus uniformly dark or black; antennae black; legs brown, with tarsi and fore tibiae testaceous; antennae not thickened distally; antennal segments in following proportions: 5:4:5:6:6:6:6:6:6:9; head dull, granulated; POL = 6; OL = 4; OOL = 4; OPL = 3; TL = 3; epistomal suture straight medially (Plate 7 A); width of malar space at its narrowest point equal to or greater than medial height of clypeus (Plate 7 A); scutum, scutellum and metanotum dull, granulated; notauli complete and posteriorly separated; often incomplete, reaching beyond 0.75 length of scutum; median mesoscutal sulcus occasionally present; propodeum reticulate rugose, with two longitudinal keels on posterior surface; median area shining, smooth, little rugose; fore wing hyaline, without dark transverse bands; ninth abdominal sternite strongly emarginated posteriorly (Fig. 24 B); distivolsella in the form of a long straight rod (Fig. 24 A); basivolsella with two subdistal bristles (Fig. 24 A).

Distribution: common in Denmark (but found only in NEZ due to under-collecting), Sweden (Sk., Bl., Ög., Nrk., Upl., Vstm.), Norway (Ø, AK, On, Bø, VE, AAy) and East Fennoscandia (Al, Ab, N, Ta). Widespread in Europe.

Biology: adults in broadleaf forests from May to August; also in pastures and fields. A parasitoid of numerous species of Typhlocybinae (Cicadellidae). Reared from *Alebra albostriella* (Fallén); *Empoasca vitis* (Göthe); *Empoasca (Kybos) smaragdula* (Fallén); *Fagocyba cruenta* (H.-S.); *Edwardsiana crataegi* (Douglas), *geometrica* (Schrank) and *lethierryi* (Edwards); *Eupterycyba jucunda* (H.-S.); *Ribautiana tenerrima* (H.-S.); *Typhlocyba quercus* (F.); *Alnetoidia alneti* (Dahlbom); *Zygina* sp. The prepupa overwinters in the soil in a cocoon (however, in England this species may also overwinter as a first instar larva in parasitized overwintering leafhoppers (Jervis, 1980b, c)). The number of generations in Fennoscandia and Denmark is not known, but in England it may be univoltine or bivoltine (Waloff and Jervis, 1987).

5. ***Aphelopus camus*** Richards, 1939
Plate 7, Fig. 25.

Aphelopus heidelbergensis Richards, 1939: 286.
Aphelopus camus Richards, 1939: 287.

Female: fully winged; length 1.68-1.81 mm; black; mandibles testaceous; clypeus whitish; antennae black or brown, with segment 1 partly yellow; legs testaceous, with mid and hind femora dark brown and with hind tibiae light brown; antennae thickened distally; antennal segments in following pro-

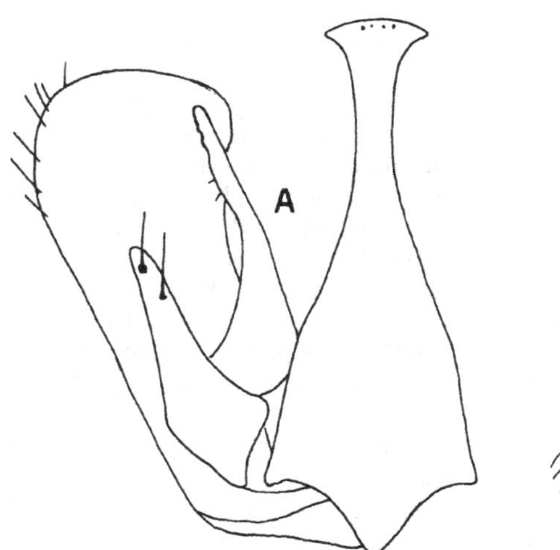

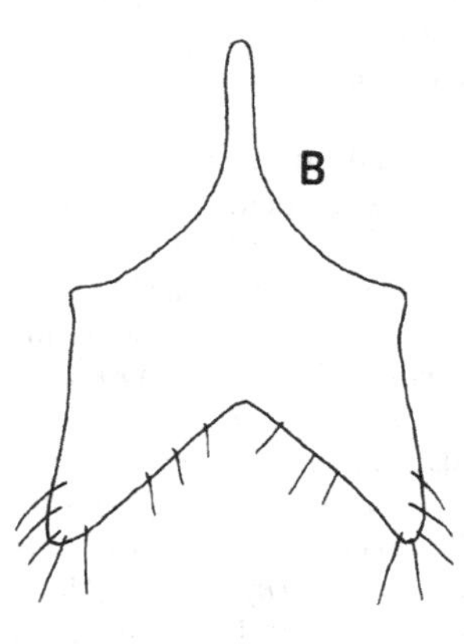

Fig. 24. Male of *Aphelopus serratus* Richards: genitalia (right half removed) (A) and 9th abdominal sternite (B).

portions: 4:3.5:5:6:7:6:6:7:6:7; head dull, granulated; frontal line incomplete; POL = 6; OL = 3; OOL = 3; OPL = 3; TL = 3; occipital carina complete; scutum, scutellum and metanotum dull, granulated; notauli incomplete, reaching approximately 0.65 length of scutum; propodeum reticulate rugose, with two longitudinal keels on posterior surface; fore wing hyaline, without dark transverse bands.

Male: fully winged; length 1.75-2.25 mm; black; mandibles and clypeus testaceous or whitish; occasionally clypeus completely brown (Plate 7 D); antennae black or brown; legs brown, with tarsi light; antennae not thickened distally; antennal segments in following proportions: 4:3:5:6:6:6:6:6:6:8; head dull, granulated; frontal line incomplete; occipital carina complete; POL = 6; OL = 4; OOL = 3; OPL = 3; TL = 3; scutum, scutellum and metanotum dull, granulated; notauli incomplete, reaching approximately 0.65 length of scutum; propodeum reticulate rugose, with two longitudinal keels on posterior surface; median area slightly less rugose than lateral regions; fore wing hyaline, without dark transverse bands; ninth abdominal sternite emarginate posteriorly (Fig. 25 B); basivolsella with 1 subdistal bristle (Fig. 25 A); aedeagus with distal apex trumpet- shaped (Fig. 25 A).

Distribution: not found in Denmark (but certainly present); rather common in Sweden (Sk., Sm., Ög., Boh., Nrk., Upl., Hls.); uncommon in Norway (On, Bø, Bv, VE); rare in East Fennoscandia, only found at Taipalsaari (Sa). Widespread in Europe, also in Turkey.

Biology: adults in broadleaf forests from June to September; also in pastures and fields. A parasitoid of Typhlocybinae (Cicadellidae). Reared only from *Chlorita viridula* (Fallén).

6. ***Aphelopus nigriceps*** Kieffer, 1905

Plate 7, Fig. 26.

Aphelopus melaleucus (Dalman) var. *nigriceps* Kieffer in Kieffer & Marshall, 1905: 219.

Female: fully winged; length 1.75-2.25 mm; black; mandibles testaceous; antennae black, with segments 1-2 testaceous; legs testaceous, with hind coxae and hind femora partly brown; antennae thickened distally; antennal segments in following proportions: 6:5.5:5:6:6:6:6:5:5:7; head dull, completely granulated; frontal line incomplete; occipital carina complete; POL = 8, OL = 4; OOL = 4; OPL = 3.5; TL = 3; scutum, scutellum and metanotum smooth, granulated; notauli incomplete, reaching approximately 0.65 length of scutum; median mesoscutal sulcus occasionally present; propodeum reticulate rugose, with two longitudi-

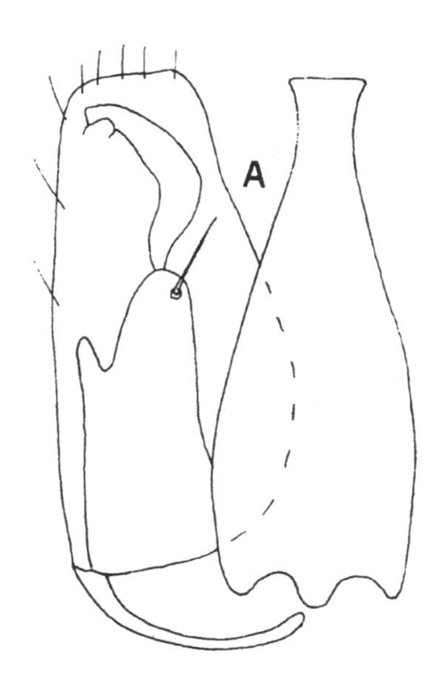

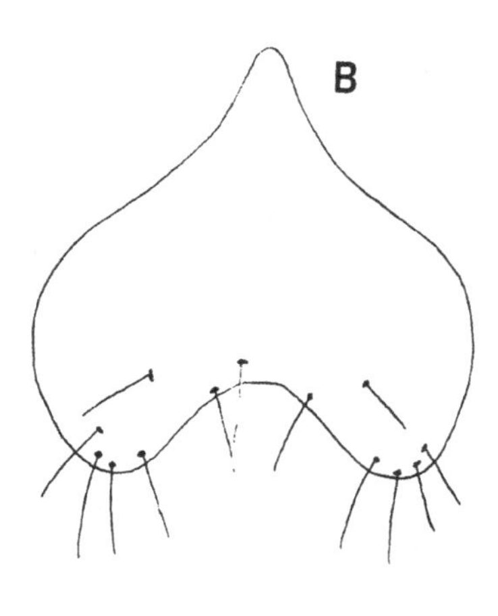

Fig. 25. Male of *Aphelopus camus* Richards: genitalia (right half removed) (A) and 9th abdominal sternite (B).

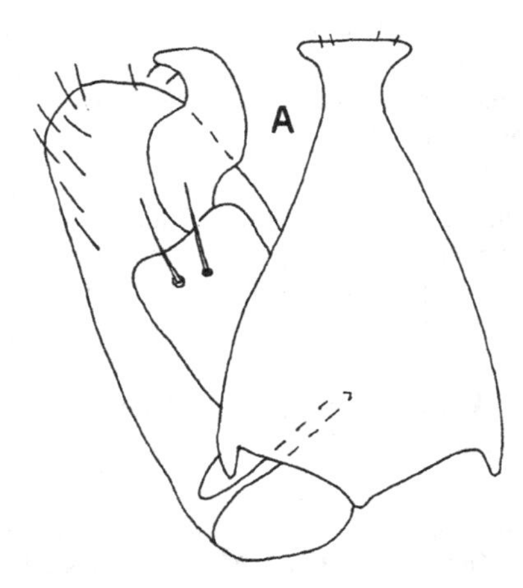

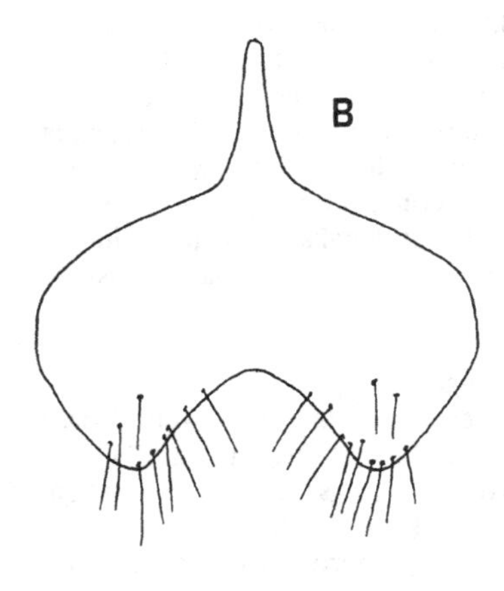

Fig. 26. Male of *Aphelopus nigriceps* Kieffer: genitalia (right half removed) (A) and 9th abdominal sternite (B).

nal keels on posterior surface; median area shining, less rugose than lateral regions; fore wing hyaline, without dark transverse bands.

Male: fully winged; length 1.75-2.37 mm; black; mandibles testaceous (Plate 7 F); antennae brown or black; legs brown, with tarsi light; antennae not thickened distally; antennal segments in following proportions: 4:4:6:6:7:7:7:8.8:11; head dull, granulated; frontal line present; occipital carina complete; POL = 9; OL = 4; OOL = 4.5; OPL = 3; TL = 4; scutum, scutellum and metanotum dull, granulated; notauli incomplete, reaching approximately 0,65 length of scutum; median mesoscutal sulcus occasionally present; propodeum reticulate rugose, with two longitudinal keels on posterior surface; median area shining, almost smooth, slightly rugose; fore wing hyaline, without dark transverse bands; ninth abdominal sternite emarginated posteriorly (Fig. 26 B); basivolsella with 2 subdistal bristles and without an outer basal process (Fig. 26 A).

Distribution: uncommon in Denmark (only found in LFM: Horreby Lyng, Falster), Sweden (Sk., Vg., Nrk., Vstm.), Norway (only found in TEi: Notodden, Lisleherad, 22.VI-6.VIII.93, by Alf Bakke) and East Fennoscandia (Al, Ab, N, Sa). Widespread, but always uncommon, in Europe, also in Nepal.

Biology: adults in broadleaf forests from May to September; also in pastures and fields. A parasitoid of Typhlocybinae (Cicadellidae). Reared only from *Empoasca vitis* (Göthe) and *Eurhadina concinna* (Germar). The prepupa overwinters in the soil in a cocoon (Jervis, 1980b). The number of generations in Fennoscandia is not known, but in England the species is univoltine and may also be bivoltine (Waloff and Jervis, 1987).

7. ***Aphelopus querceus*** Olmi, 1984

Plate 7, Fig. 27.

Aphelopus querceus Olmi, 1984: 59.

Female: fully winged; length 1.81-2.31 mm; black; head with mandibles, clypeus, lower face and a U-shaped mark (enclosing bases of antennae, with arms lying for a short distance along inner margin of eyes) whitish; often head without U-shaped mark, but with only lower face whitish (Plate 7 G) or with entire frons from anterior ocellus to clypeus whitish; antennae black, with segments 1-2 testaceous; legs whitish or testaceous; antennae slightly thickened distally; antennal segments in following proportions: 6:4:5.5:6:7:7:6:5:5:8; head dull, granulated; frontal line complete; occipital carina complete; POL = 5.5; OL = 3.5; OOL = 3; OPL = 5; TL = 5; scutum dull, granulated; notauli incomplete, reaching approximately 0.60-0.75 length of scutum; scutellum and metanotum shining, smooth, without sculpture; propodeum reticulate rugose; fore wing hyaline, without dark transverse bands.

Male: fully winged; length 1.81-2.18 mm; black; mandibles and clypeus testaceous; head black, usually with some whitish or testaceous or brown spots between antennal sockets (Plate 7 E); malar space more or less whitish or testaceous;

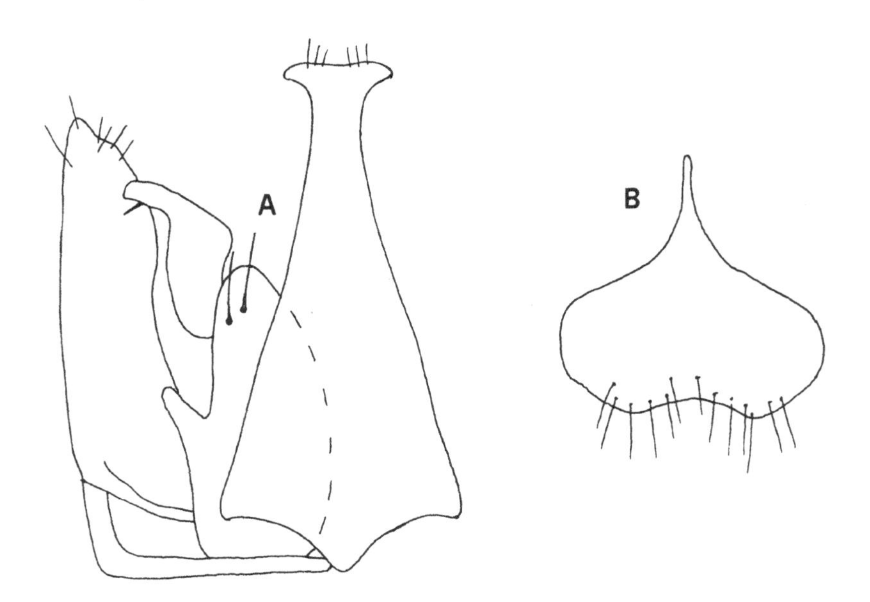

Fig. 27. Male of *Aphelopus querceus* Olmi: genitalia (right half removed) (A) and 9th abdominal sternite (B).

antennae black; legs testaceous, with hind femora darkened; antennae not thickened distally; antennal segments in following proportions: 5:5:7:8:9:9:9:9.5:9.5:14; head dull, granulated; frontal line complete; occipital carina complete; POL = 6; OL = 3.5; OOL = 5; OPL = 2.5; TL = 2.5; scutum dull, granulated; notauli incomplete, reaching approximately 0.60-0.75 length of scutum; scutellum granulated; metanotum shining, smooth, without sculpture; propodeum reticulate rugose, without transverse or longitudinal keels, with a shining and smooth median area on the posterior surface; fore wing hyaline, without dark transverse bands; ninth abdominal sternite weakly emarginate posteriorly (Fig. 27 B); basivolsella with 2 subdistal bristles (Fig. 27 A); distal apex of aedeagus not tridentate (Fig. 27 A).

Distribution: not found in Denmark; uncommon in Sweden (Nrk., Vstm.), Norway (Ø, AK, Bø, VE) and East Fennoscandia (Ab, Sa). Uncommon in other European countries; known only from Netherlands, France, Italy, England; also in Nepal.

Biology: adults in broadleaf forests in July and August; also in pastures and fields. A parasitoid of Typhlocybinae (Cicadellidae). Reared only from *Empoasca vitis* (Göthe) and *solani* (Curtis).

2. Subfamily Anteoninae

Type genus: ***Anteon*** Jurine, 1807.

Female: fully winged; rarely brachypterous; fore wing with costal, basal and subbasal cells completely enclosed by pigmented veins; maxillary palps 6-segmented; labial palps 3-segmented; front tarsus chelate; chela without rudimentary claw; enlarged claw without teeth and lamellae; mandibles quadridentate; pronotal tubercles present; occipital carina complete; antennae without tufts of long hairs; tibial spurs 1, 1, 2.

Male: fully winged; fore wing with costal, basal and subbasal cells completely enclosed by pigmented veins; maxillary palps 6- segmented; labial palps 3-segmented; mandibles quadridentate; occipital carina complete; parameres without a dorsal process; fore wing with marginal vein shorter than pterostigma; tibial spurs 1, 1, 2.

Distribution: worldwide.

Hosts: Cicadellidae (except for Typhlocybinae).

Genera: 2 genera in Fennoscandia and Denmark.

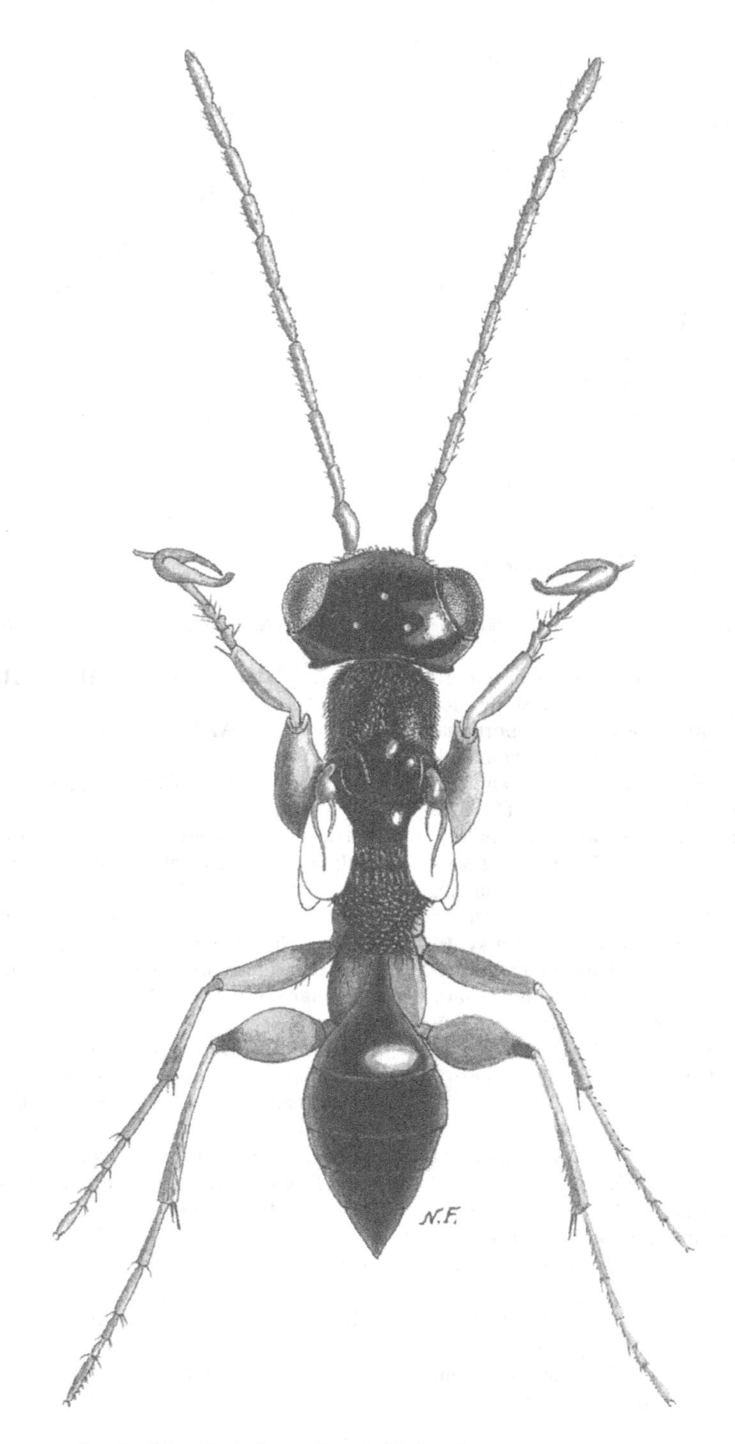

Fig. 28. Micropterous female of *Lonchodryinus ruficornis* (Dalman).

2. Genus *Lonchodryinus* Kieffer, 1905

Lonchodryinus Kieffer, 1905: 95.
  Type-species: *Lonchodryinus tricolor* Kieffer, 1905, by monotypy.
Prenanteon Kieffer, 1913: 301.
  Type-species: *Anteon melanocera* Kieffer, 1905, by subsequent designation.
Psilanteon Kieffer, 1913: 301.
  Type species: *Anteon aequalis* Kieffer, 1905, by original designation.

Female: fully winged or, occasionally, brachypterous; fore wing with distal part of stigmal vein as long as or longer than proximal part, usually forming a curve between proximal and distal parts (Plate 9); occasionally distal part slightly shorter than proximal part; propodeum usually without a strong transverse keel between dorsal and posterior surfaces.

Male: very different from female; prothorax more reduced than in the opposite sex; fully winged; fore wing with distal part of stigmal vein as long as or longer than proximal part, usually forming a curve between proximal and distal parts (Plate 11); occasionally distal part slightly shorter than proximal part; propodeum usually without a strong transverse keel between dorsal and posterior surfaces.

Distribution: worldwide.

Hosts: Cicadellidae (except for Typhlocybinae).

Species: one species in Fennoscandia and Denmark.

8. *Lonchodryinus ruficornis* (Dalman, 1818)

Plates 9-11, Figs 28-30.

Gonatopus ruficornis Dalman, 1818: 83.
Gonatopus frontalis Dalman, 1818: 84.
Gonatopus basalis Dalman, 1818: 84.
Gonatopus fuscicornis Dalman, 1818: 87.
Dryinus longicornis Dalman, 1823: 10.
Dryinus daos Walker, 1837: 423.
Dryinus lapponicus Thomson, 1860: 176.
Dryinus retusus Thomson, 1860: 176.
Anteon subapterus Kieffer in Kieffer & Marshall, 1905: 138.
Anteon vitellinipes Kieffer in Kieffer & Marshall, 1905: 181.

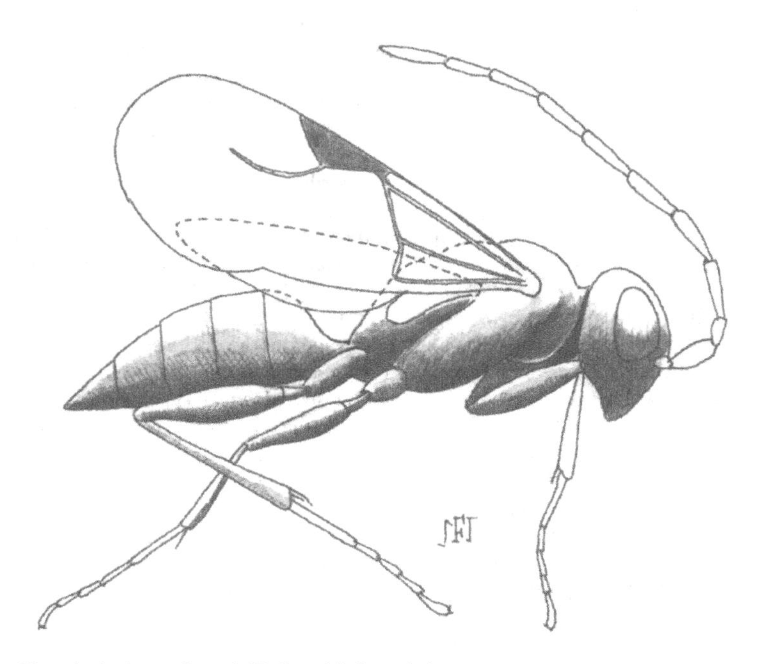

Fig. 29. Male of *Lonchodryinus ruficornis* (Dalman) in lateral view.

Anteon försteri Kieffer in Kieffer & Marshall, 1905: 183.
Anteon dolichocerus Kieffer in Kieffer & Marshall, 1905: 190.
Anteon longicornis (Dalman) ab. *medialis* Hellén, 1919: 284.
Anteon longicornis (Dalman) ab. *nigrofemoratus* Hellén, 1919: 284.
Anteon longicornis (Dalman) ab. *aterrimus* Hellén, 1919: 284.
Prenanteon borealis Hellén, 1935: 7.

Female (Plates 9-10, Fig. 28): fully winged; occasionally micropterous (Fig. 28); rarely brachypterous; length 1.87-4.68 mm; black; occasionally head and thorax reddish-brown; mandibles testaceous, with teeth dark; occasionally frons reddish; occasionally malar space and frons mostly whitish or yellow (Plate 10); clypeus yellow, or black, or dark reddish; antennae completely testaceous; occasionally brown, with segments 1-2 or 1-3 or only 1 testaceous; occasionally antennae brown, with segments 6-10 or 7-10 testaceous; tegulae testaceous; legs completely testaceous or with hind coxae basally black; gaster black or brown-testaceous; antennae thickened distally; antennal segments in following proportions: 12:6:13:14:12:12:10:10:10:15; head shining, strongly punctate, smooth, without sculpture among punctures; occipital carina complete; POL = 7; OL = 4; OOL = 10; OPL = 8; TL = 8; frons usually without a median groove, with only a discal impression; occasionally with a deep median longitudinal groove; this frontal impression very variable, but of no systematic value; pronotum shining, strongly punctate; pronotal tubercles reaching tegulae; scutum shining, smooth, without punctures or finely or strongly punctate, without sculpture among punctures; notauli incomplete; very variable both in length, from 0.3 to 0.75 length of scutum, and in appearance: very thin or strong or aciculate; scutellum and metanotum shining, smooth, without sculpture or punctate; propodeum reticulate rugose, without a strong transverse keel between dorsal and posterior surfaces; posterior surface with two longitudinal keels; median area usually as rugose as lateral regions; occasionally median area shining, smooth, not rugose; fore wing hyaline, without dark transverse bands; distal part of stigmal vein as long as, or slightly longer than, or slightly shorter than, proximal part; stigmal vein forming a curve between proximal and distal parts; macropterous specimens very common; micropterous forms and intermediate brachypterous specimens uncommon; fore tarsal segment 1 slightly longer than segment 4 (9:7); segment 2 of fore tarsus produced into a hook; enlarged claw (Fig. 30 A) with a proximal prominence bearing a long bristle; segment 5 of fore tarsus (Fig. 30 A) with 1 row of approximately 15-20 lamellae; distal apex with a group of approximately 3-7 lamellae.

Male (Plate 11, Fig. 29): fully winged; length 2.12-3.75 mm; black; mandibles testaceous, with teeth brown; antennae wholly brown, occasionally testaceous; tegulae testaceous; legs testaceous, with coxae partly brown; occasionally clubs of femora partly brown; head shining, finely punctate, without sculpture among punctures; occipital carina complete; POL = 5; OL = 2.5; OOL = 8; OPL = 4; TL = 5; frons with or without a more or less deep central impression; scutum shining, smooth, finely punctate, without sculpture among punctures; notauli incomplete, reaching approximately 0.5 length of scutum; notauli very variable in appearance, either deep and broad or narrow and finely crenulate; scutellum and metanotum shining, smooth, finely punctate, without sculpture among punctures; propodeum reticulate rugose, without a strong transverse keel between dorsal and posterior surfaces; posterior surface with two longitudinal keels; median area usually as rugose as lateral regions; occasionally smooth and shining, almost without sculpture; fore wing hyaline, without dark transverse bands; distal part of stigmal vein as long as, slightly shorter than, or slightly longer than, proximal part; stigmal vein forming a curve between proximal and distal parts; parameres (Fig. 30 B) without an inner distal pointed process; their lateral shape very variable, from narrow to broad.

Distribution: common in Denmark (SJ, EJ, WJ, NWJ, F, LFM, NWZ, NEZ), Sweden (abundant throughout the entire country, and not only found in Hall., G. Sand., Boh., Dlsl., Gstr., Ly. Lpm.), Norway (AK, Os, On, Bø, Bv, VE, AAy, NTi, Nsi, Nsy, Fv) and East Fennoscandia (abundant throughout the entire country, and not only found in Ka, Oa, Om, ObS, LkE). Gunnarnes (71°, Rolvsøy, Måsøy, Fv) is the most northerly known occurrence of Dryinidae anywhere in the world. Widespread in Europe, also in Japan, Korea, Mongolia, Nepal, Siberia, Kazakhstan.

Biology: adults in broadleaf forests, pastures and fields from April to September. A parasitoid of Cicadellidae Deltocephalinae. Reared from *Macrosteles laevis* (Ribaut); *Elymana sulphurella* (Zetterstedt); *Conosanus obsoletus* (Kirschbaum); *Euscelis incisus* (Kirschbaum); *Streptanus sordidus*

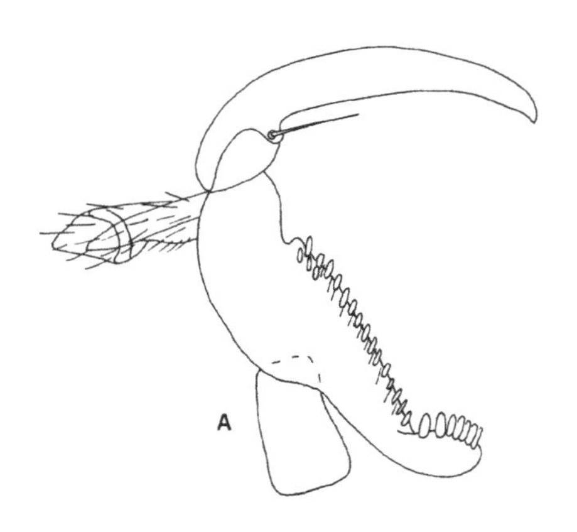

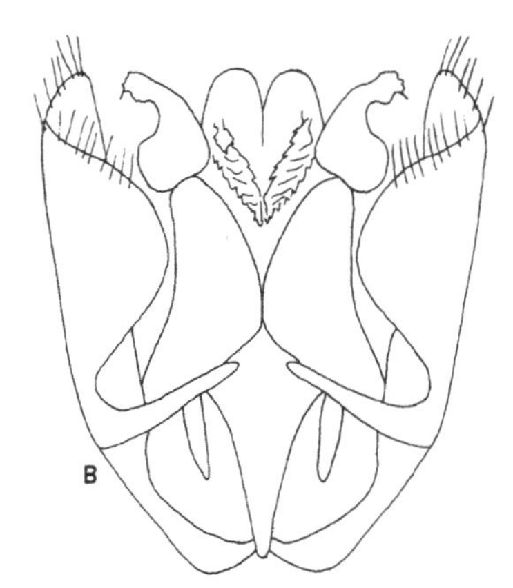

Fig. 30. *Lonchodryinus ruficornis* (Dalman): chela (A) and male genitalia (B).

(Zetterstedt); *Psammotettix cephalotes* (H.-S.), *confinis* (Dahlbom) and *nodosus* (Ribaut); *Errastunus ocellaris* (Fallén); *Jassargus distinguendus* (Flor) and *flori* (Fieber); *Arthaldeus pascuellus* (Fallén). The mature larva overwinters in a cocoon spun on leaves (Waloff, 1974). The number of generations in Fennoscandia and Denmark is not known, but in England the species may be bivoltine or univoltine (Waloff, 1974, 1975). It may be parasitized by *Ismarus rugulosus* Förster (Waloff, 1975, misid. *Prenanteon* sp.).

3. Genus *Anteon* Jurine, 1807

Anteon Jurine, 1807: 302.
  Type-species: *Anteon jurineanum* Latreille, 1809, first included species.
Chelogynus Haliday, 1838: 518.
  Type-species: *Dryinus infectus* Haliday, 1837, by subsequent designation.
Neochelogynus Perkins, 1905: 60.
  Type-species: *Neochelogynus typicus* Perkins, 1905, by original designation.

Female: fully winged; fore wing with distal part of stigmal vein much shorter than proximal part, forming an angle between proximal and distal parts (Plates 12-15, Fig. 31); occasionally distal part slightly shorter, as long as, or slightly longer than, proximal part; propodeum with a strong transverse keel between dorsal and posterior surfaces.

Male: very different from female; prothorax more reduced than in the opposite sex (Plates 16-17); other characteristics as in above female description.

Distribution: worldwide.

Hosts: Cicadellidae (except for Typhlocybinae).

Species: 11 species in Fennoscandia and Denmark.

Key to species of *Anteon*

Females

1 Segment 4 of fore tarsus at most 0.5 as long as segment 1; segment 5 of fore tarsus with basal part longer than distal part (Fig. 32 B); fore tarsal segment 3 or 4 produced into a hook 2
– Segment 4 of fore tarsus at least 0.66 as long as segment 1; segment 5 of fore tarsus with basal part as long as or shorter than distal part (Fig. 38); fore tarsal segment 2 produced into a hook 5
2 Posterior surface of propodeum without longitudinal keels ..... 9. *jurineanum* Latreille
– Posterior surface of propodeum with two longitudinal keels (Plates 12-15) 3
3 Frons not carinate centrally; occasionally with a short median keel visible near anterior ocellus; antennae black .... 10. *brachycerum* (Dalman)
– Frons carinate centrally; median keel complete or reaching at least mid-length of frons; median keel rarely absent or reduced to a trace, but then antennae wholly yellow ... 4
4 Antennae at least partly black or brown ... . 11. *arcuatum* Kieffer

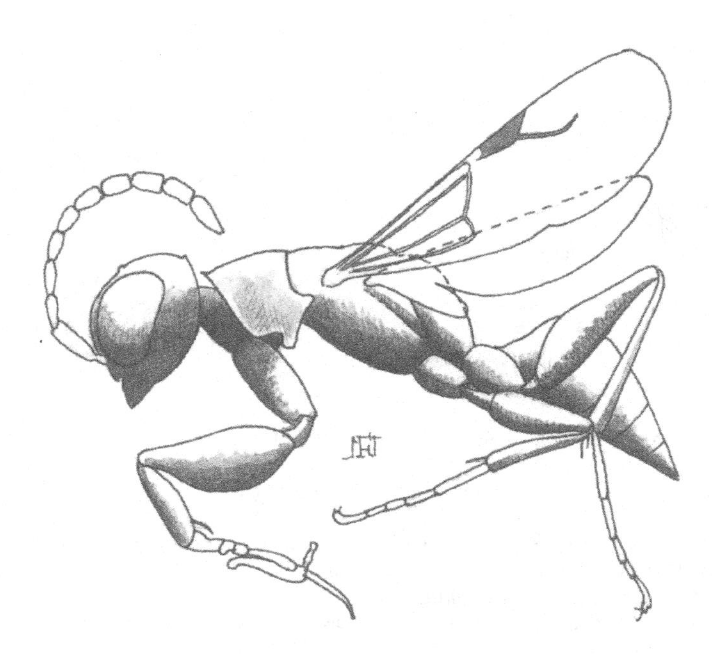

Fig. 31. Female of *Anteon gaullei* Kieffer in lateral view.

– Antennae wholly yellow .... 12. *flavicorne* (Dalman)

5 Posterior surface of propodeum with median area shining, mostly smooth and without sculpture, only partly rugose 6

– Posterior surface of propodeum with median area dull, completely rugose, approximately as rugose as lateral regions 8

6 Head, pronotum, scutum and scutellum yellow or reddish .... 13. *ephippiger* (Dalman)

– Head and thorax completely black; at most clypeus and frons partly testaceous 7

7 Species very small, less than 2.5 mm long; head finely punctate and completely smooth; frons smooth, without keels or areolae; segment 5 of fore tarsus with 1 row of lamellae (Fig. 37 B) .... 14. *pubicorne* (Dalman)

– Species large, more than 3.0 mm long; head strongly punctate; frons with keels or areolae; segment 5 of fore tarsus with two rows of lamellae (Fig. 38 B) .. 15. *infectum* (Haliday)

8 Pronotum with posterior surface transverse, much broader than long (Fig. 39 C) .... 16. *exiguum* (Haupt)

– Pronotum not transverse, with posterior surface approximately as long as broad (Fig. 39 D) 9

9 Ocellar triangle delimited by keels joining ocelli; lateral keels occasionally slightly visible; posterior keel always visible; head and thorax black .... 17. *tripartitum* Kieffer

– Ocellar triangle not delimited by keels; head and thorax black or differently coloured 10

10 Head, thorax and propodeum completely black, occasionally brown or with cupreous tinges; head punctate, without sculpture among punctures .. 14. *pubicorne* (Dalman)

– Head, thorax and propodeum partly or completely yellow, or testaceous, or reddish, or dark-reddish; occasionally completely dark-brown, or almost wholly or wholly black, but then head granulated among punctures .. 11

11 Head wholly black; scutum completely or mainly black, pronotum reddish, or yellow, or testaceous; segment 5 of fore tarsus with two rows of lamellae (Fig. 41 B) .... 18. *gaullei* Kieffer

– At least head partly yellow, or reddish, or testaceous; occasionally head and scutum dark brown or black, but then pronotum dark brown, or black, or blackish; segment 5 of fore tarsus with 1 row of lamellae (Figs 36 B, 19) 12

12 Head dull, always clearly and strongly granulated; antennal segment 1 approximately twice as long as segment 4 .... 19. *fulviventre* (Haliday)

– Head shining, punctate, without sculpture among punctures or very weakly granulated; antennal segment 1 approximately as long as or slightly longer than segment 4 . . . . 13. *ephippiger* (Dalman)

Males

1 Posterior surface of propodeum without longitudinal keels 2
– Posterior surface of propodeum with two longitudinal keels (Plates 16-17) 3
2 Head punctate, without sculpture among punctures; parameres with a distal inner pointed process (Fig. 37 A) . . . . 14. *pubicorne* (Dalman)
– Head more or less rugose; parameres with no distal inner pointed process (Fig. 32 A) . . . . 9. *jurineanum* Latreille
3 Parameres with no distal inner more or less pointed process (Figs 32 A, 38 A) 4
– Parameres with a distal inner more or less pointed process (Figs 34 A, 39 A) 5
4 Posterior surface of propodeum with median area shining and almost completely smooth, not rugose . . . . 15. *infectum* (Haliday)
– Posterior surface of propodeum with median area dull and rugose . . . . 9. *jurineanum* Latreille
5 Parameres with no dorsal membranous process (Fig. 33 A); head granulated, not punctate; antennae and legs mostly brown or black . . . . 10. *brachycerum* (Dalman)
– Parameres with a more or less large dorsal membranous process (Figs 34 A, 39 A); antennae and legs testaceous, or brown, or black 6
6 Distal inner process of parameres extended medially and with inner margin excavated (Figs 34 A, 35 A) 7
– Distal inner process of parameres extended apically and with inner margin convex or straight (Figs 36 A, 37 A) 8
7 Legs testaceous; at most hind legs with coxae, femora and tibiae partly brown . . . . 12. *flavicorne* (Dalman)
– Legs testaceous, with coxae and clubs of femora brown or dark . . . . 11. *arcuatum* Kieffer
8 Legs completely yellow or testaceous, occasionally with proximal extremities of hind coxae brown; occasionally stalks of hind femora brown 9
– Legs more or less brown or blackish, with at least mid and hind coxae and femora brown or blackish 13
9 Head strongly granulated, smooth, not punctate; rarely with irregular keels 10
– Head strongly or finely punctate, without sculpture among punctures; occasionally weakly granulated, or alutaceous, or partly rugose, but then always with visible punctures 11
10 Head smooth, completely granulated . . . . 19. *fulviventre* (Haliday)
– Head granulated and sculptured by irregular keels, not smooth . . . 17. *tripartitum* Kieffer
11 Proximal dorsal membranous process of parameres very short (Fig. 39 A); head in part weakly granulated, in part rugose, in part punctate, occasionally alutaceous, with sculpture usually weakly distinct . . . . 16. *exiguum* (Haupt)
– Proximal dorsal membranous process of the parameres very long (Figs 36 A, 41 A); head punctate, without sculpture among punctures; rarely weakly granulated among punctures; head surface never alutaceous 12
12 Head finely punctate, smooth, without sculpture among punctures or very weakly granulated, usually with no median frontal line 13. *ephippiger* (Dalman)
– Head more strongly punctate, without sculpture among punctures or very weakly granulated, with a short or long median frontal line . . . . 18. *gaullei* Kieffer
13 Head strongly granulated, occasionally with areolae and irregular keels . . . . 17. *tripartitum* Kieffer
– Head punctate, without sculpture among punctures or very weakly granulated . . . . 14. *pubicorne* (Dalman)

9. *Anteon jurineanum* Latreille, 1809

Fig. 32.

Anteon jurineanum Latreille, 1809: 35.
Gonatopus brevicornis Dalman, 1818: 85.
Anteon crenulatus Kieffer in Kieffer & Marshall, 1905:141.
Anteon thomsoni Kieffer in Kieffer & Marshall, 1905: 142.
Anteon vicinus Kieffer in Kieffer & Marshall, 1905: 145.
Anteon marginatus Kieffer in Kieffer & Marshall, 1905: 147.
Antaeon barbatus Chitty, 1908: 142.

Female: fully winged; length 2.00-3.12 mm; black; mandibles testaceous, with brown teeth; antennae wholly brown or with segments 1-2 or 1-5 testaceous; legs brown, with fore tibiae and tarsi light; occasionally legs testaceous, with hind coxae partly brown, or with coxae, clubs of femora and tibiae partly brown or blackish; tegulae testaceous; antennae short, thickened distally; antennal segments in following proportions: 12:4:5:5:5:5:4:5:5:8; head dull, granulated, with more or less numerous areolae or irregular keels; face with a frontal line, with or without lateral keels near orbits directed towards antennal sockets; occasionally these lateral keels incomplete; occipital carina complete; POL, OL, OOL, OPL and TL very variable; pronotum dull, rugose, with posterior surface short, shorter than scutum (4:18); pronotal tubercles reaching tegulae; scutum dull, granulated and reticulate rugose, occasionally only granulated; notauli incomplete, reaching approximately mid-length of scutum or shorter; scutellum and metanotum shining, smooth and without sculpture, or rugose; propodeum reticulate rugose, with a strong transverse keel between dorsal and posterior surfaces; posterior surface without longitudinal keels; fore wing hyaline, without dark transverse bands; distal part of stigmal vein much shorter than proximal part (3:8); fore tarsal segments in following proportions: 9:2:2:2:8; segment 4 of fore tarsus produced into a hook; enlarged claw (Fig. 32 B) with a proximal prominence bearing a long bristle; segment 5 of fore tarsus (Fig. 32 B) with 1 row of 1-4 lamellae, as well as numerous bristles; distal apex with a group of 1-3 lamellae.

Male: fully winged; length 1.56-3.00 mm; black; mandibles testaceous; antennae brown or black; legs more or less brown, with tarsi and fore tibiae light; tegulae testaceous; antennae not thickened distally; antennal segments in following proportions: 12:6:7:8:8:8:8:8:7:12; head dull, granulated, with more or less numerous areolae or irregular keels; face with a frontal line, with or without lateral keels near orbits directed towards antennal sockets; occasionally these lateral keels incomplete; POL = 8; OL = 4; OOL = 7; OPL = 3; TL = 4; occipital carina complete; scutum dull or shining, smooth, without sculpture, occasionally weakly or strongly granulated; anterior surface of scutum more strongly sculptured; notauli incomplete, reaching mid-length of scutum or shorter; scutellum and metanotum shining, smooth, without sculpture; propodeum reticulate rugose, with a strong transverse keel between dorsal and posterior surfaces; posterior surface usually without longitudinal keels, occasionally with two distinct longitudinal keels and in this case median area as rugose as lateral regions; fore wing hyaline, without dark transverse bands; distal part of stigmal vein

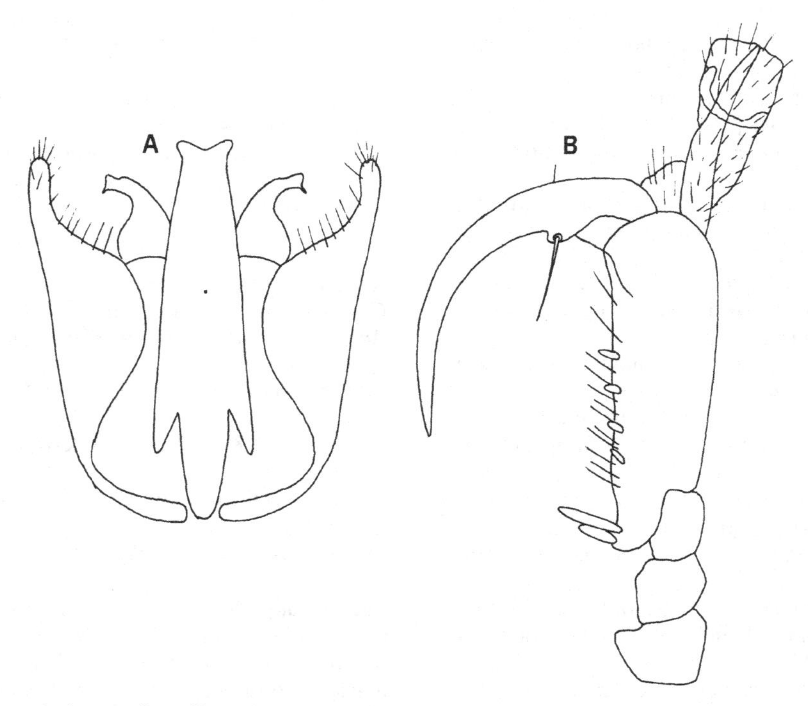

Fig. 32. *Anteon jurineanum* Latreille: male genitalia (A) and chela (B).

much shorter than proximal part (3:9); parameres (Fig. 32 A) without a distal inner pointed process and without a membranous proximal process; parameres, in lateral view, very variable in shape; occasionally with apical region very long and narrow, or broad and pointed, or rounded, or with intermediate characteristics.

Distribution: common in Denmark (SJ, EJ, F, LFM, SZ, NEZ), Sweden (Sk., Bl., Sm., Öl., Ög., Vg., Nrk., Sdm., Upl., Vstm., Vrm., Hls.), Norway (Bø, Bv, TEi, Nnø) and East Fennoscandia (Al, Ab, N, Ka, St, Ta, Sa, Oa, Tb, Kb, Ok, Ks, LkW, Li, Vib, Kr). Widespread in Europe, also in Japan, Mongolia, Nepal, Siberia, Turkey, Jordan.

Biology: adults in broadleaf forests, pastures and fields from April to July. A parasitoid of Cicadellidae Macropsinae. Reared from *Oncopsis flavicollis* (L.). This species may be parasitized by *Ismarus halidayi* Förster (Hymenoptera Diapriidae) (Chambers, 1955, misid. *Anteon brevicorne* (Dalman)).

10. ***Anteon brachycerum*** (Dalman, 1823)
Fig. 33.

Dryinus brachycerus Dalman, 1823: 12.
Anteon obscuricornis Kieffer in Kieffer & Marshall, 1906: 523.

Female: fully winged; length 2.30-3.30 mm; black, with yellow mandibles (except for brown teeth) and brown legs (with light tarsi and joints); tegulae testaceous; antennae short, with segment 1 approximately twice as long as 4; head dull, smooth, granulated; frontal line absent; occasionally with a short frontal line visible near anterior ocellus; scutum shining, punctate, smooth; notauli incomplete, reaching approximately mid- length of scutum; scutellum and metanotum shining, smooth, without sculpture; propodeum reticulate rugose, with a strong transverse keel between dorsal and posterior surfaces; posterior surface with two longitudinal keels; median area very shining (more than lateral regions), almost smooth or weakly rugose; fore wing hyaline, without dark transverse bands; distal part of stigmal vein much shorter than proximal part; segment 1 of fore tarsus more than twice as long as segment 4; fore tarsal segment 3 produced into a hook; enlarged claw (Fig. 33 B) with a proximal prominence bearing a long bristle; segment 5 of fore tarsus (Fig. 33 B) with basal part longer than distal part, with approximately 10-13 bristles; distal apex with a group of 4 lamellae.

Male: fully winged; length 2.30-3.30 mm; black, with brown antennae and yellow mandibles (except for brown teeth); legs completely brown or with light tarsi and joints; tegulae testaceous; head smooth, dull, granulated, not punctate; frontal line absent; scutum shining, haired, punctate, smooth; notauli incomplete, reaching approximately midlength of scutum; scutellum and metanotum shining, without sculpture; propodeum reticulate rugose, with a strong transverse keel between dorsal and posterior surfaces; posterior surface with two longitudinal keels; median area very shining (more than lateral regions), smooth; fore wing hyaline, without dark transverse bands; parameres (Fig. 33 A) with a distal inner pointed process, without proximal dorsal membranous process.

Distribution: common in Denmark (SJ, EJ, NWZ, NEZ), Sweden (Sk., Bl., Sm., Öl., Ög., Nrk., Upl., Gstr., Vb.), Norway (AK, TEy, HOy) and East Fennoscandia (Al, Ab, N, Ta, Sa, Tb, Kr). Widespread in Central, Northern and Eastern Europe; not found in Southern Europe; also in Nepal.

Biology: adults in broadleaf forests from April to June. A parasitoid of Cicadellidae Macropsinae. Reared from *Oncopsis flavicollis* (L.). Mating observed by Jervis (1979b).

11. ***Anteon arcuatum*** Kieffer, 1905
Fig. 34.

Anteon arcuatum Kieffer in Kieffer & Marshall, 1905: 144.
Anteon flavicorne (Dalman) var. *bensoni* Richards, 1939: 255.

Female: fully winged; length 2.18-3.00 mm; black; antennae brown, occasionally partly testaceous; legs testaceous, with coxae partly brown and clubs of hind femora brown; tegulae testaceous; mandibles yellow, with brown teeth; antennae short, with segment 1 approximately twice as long as segment 4; head shining, smooth, without sculpture or weakly alutaceous, with irregular keels at least near occipital triangle; frons with a frontal line and two lateral keels near orbits; pronotum rugose; scutum smooth, shining, with anterior half punctate and posterior half almost without punctures; notauli incomplete, reaching approximately midlength of scutum; scutellum and metanotum shining, smooth, without sculpture; propodeum reticulate rugose, with a transverse keel between dorsal and posterior surface; posterior surface with two longitudinal keels; median area usually dull, ru-

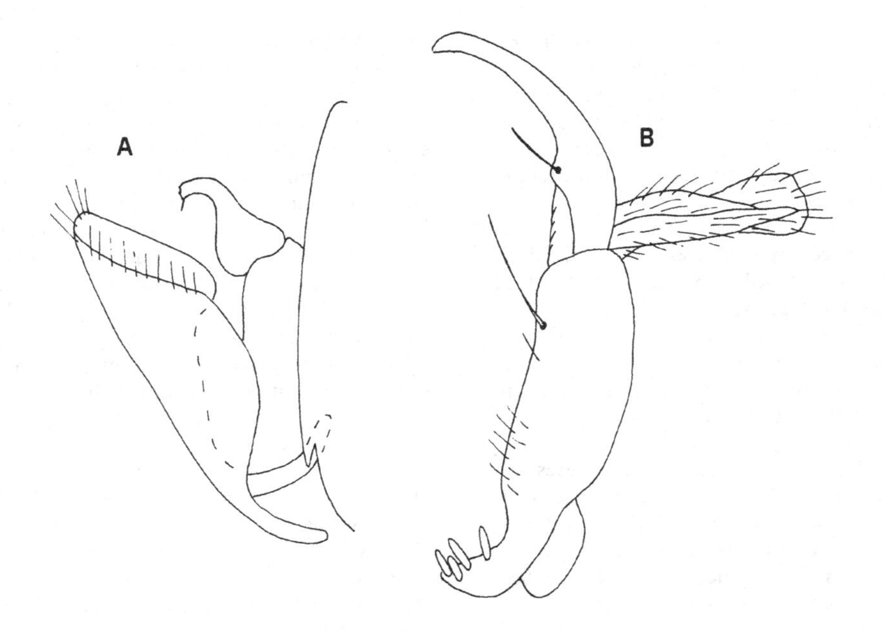

Fig. 33. *Anteon brachycerum* (Dalman): male genitalia (right half removed) (A) and chela (B).

gose, approximately as rugose as lateral regions; occasionally median area smooth; fore wing hyaline, without dark transverse bands; distal part of stigmal vein much shorter than proximal part; segment 1 of fore tarsus more than twice as long as segment 4; fore tarsal segment 3 produced into a hook; enlarged claw (Fig. 34 B) with a proximal prominence bearing a long bristle; segment 5 of fore tarsus (Fig. 34 B) with basal part longer than distal part, with 1 row of 8-18 bristles; distal apex with a group of 4 lamellae.

Male: fully winged; length 2.31-2.50 mm; black; antennae testaceous, with dorsal part of segments brown; occasionally only segments 1-3 brown; mandibles yellow, with brown teeth; legs testaceous, with brown coxae and femora; occasionally only hind femora brown; tegulae testaceous; head dull or shining, granulated, with numerous areolae or irregular keels; frontal line present; frons without lateral longitudinal keels; scutum shining, smooth, without sculpture, with anterior half punctate and posterior half almost without punctures; notauli incomplete, reaching approximately mid-length of scutum; scutellum and metanotum shining, smooth, without sculpture; propodeum reticulate rugose, with a strong transverse keel between dorsal and posterior surface; posterior surface with two longitudinal keels; median area shining, less rugose than lateral regions; fore wing hyaline, without dark transverse bands; distal part of stigmal vein much shorter than proximal part; parameres (Fig. 34 A) with a distal inner process pointed and extended medially; inner margin of this process excavated; dorsal proximal membranous process present.

Distribution: rather common in Denmark (EJ, NWJ, F, NEZ), Sweden (Sk., Bl., Ög., Vg., Nrk., Upl., Vstm., Vb.), Norway (AK) and East Fennoscandia (N, St, Sa, Oa, Tb, Kb, Kr). Widespread in Europe, also in Mongolia and Siberia.

Biology: adults in broadleaf forests from May to July. A parasitoid of Cicadellidae Idiocerinae. Reared from *Rhytidodus decimusquartus* (Schrank); *Idiocerus* sp. This species may be parasitized by *Ismarus flavicornis* (Thomson) (Hymenoptera Diapriidae) (Chambers, 1955, misid. *Anteon flavicorne* (Dalman) var. *bensoni* Richards).

12. ***Anteon flavicorne*** (Dalman, 1818)
Fig. 35.

Gonatopus flavicornis Dalman, 1818: 83.

Female: fully winged; length 2.18-4.00 mm; black; antennae yellow, occasionally with fuscous dorsal markings; legs yellow, with hind coxae partly black and hind femora with brown stalk; tegulae testa-

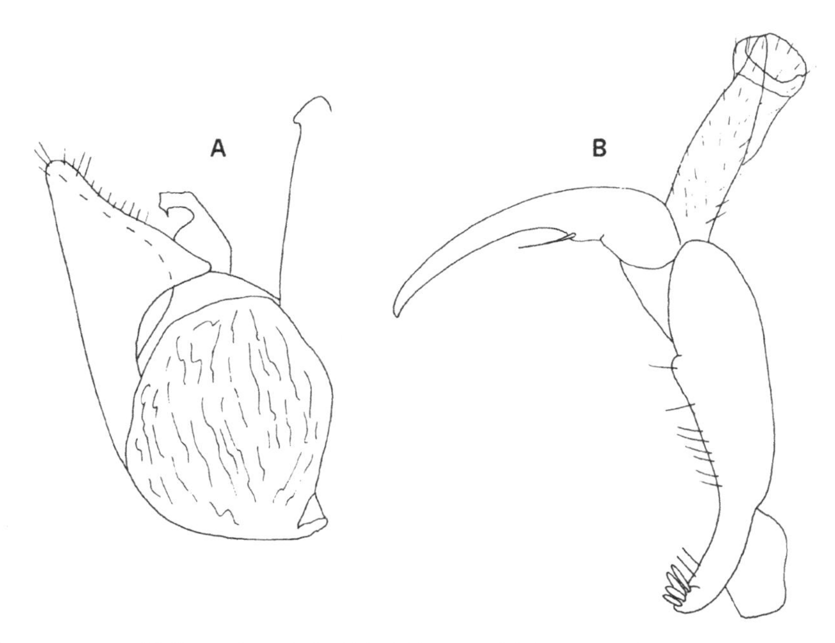

Fig. 34. *Anteon arcuatum* Kieffer: male genitalia (right half removed) (A) and chela (B).

ceous; mandibles yellow, with brown teeth; antennae short, with segment 1 approximately twice as long as segment 4; head shining, punctate, strongly or weakly granulated, with a few irregular keels near occipital carina; frontal line present, usually complete, occasionally incomplete or reduced to a trace; frons with two lateral complete or incomplete keels situated near orbits and directed towards antennal sockets; scutum shining, smooth, with anterior third strongly punctate and with posterior surface without punctures; notauli incomplete, reaching approximately 0.3 length of scutum; scutellum and metanotum shining, smooth, without sculpture; propodeum reticulate rugose, with a transverse keel between dorsal and posterior surfaces; posterior surface with two longitudinal keels; median area smooth, rugose near the margins, projecting or not forward beyond anterior margin of lateral areas; fore wing hyaline, without dark transverse bands; distal part of stigmal vein much shorter than proximal part; segment 1 of fore tarsus more than twice as long as segment 4; fore tarsal segment 3 produced into a hook; enlarged claw (Fig. 35 B) with a proximal prominence bearing a long bristle; segment 5 of fore tarsus (Fig. 35 B) with basal part longer than distal part, with 1 row of numerous bristles; distal apex with a group of 8 lamellae.

Male: fully winged; length 2.00-2.62 mm; black; antennae testaceous, with fuscous dorsal markings; legs completely testaceous; at most hind legs with coxae, tibiae and clubs of femora partly brown; tegulae testaceous; mandibles yellow, with brown teeth; head dull, granulated, wholly or partly reticulate rugose and punctate; scutum shining, smooth, finely punctate; notauli incomplete, reaching approximately 0.3 length of scutum; scutellum and metanotum shining, smooth, finely punctate; propodeum reticulate rugose, with a strong transverse keel between dorsal and posterior surfaces; median area shining, almost smooth, partly rugose; fore wing hyaline, without dark transverse bands; distal part of stigmal vein much shorter than proximal part; parameres (Fig. 35 A) with a distal inner process pointed and extended medially; inner margin of this process excavated; dorsal proximal membranous process present.

Distribution: common in Denmark (SJ, SZ, NEZ), Sweden (Sk., Sm., Vg., Upl., Vstm., Vrm.), Norway (Ø, AK, TEi) and East Fennoscandia (Ab, N, St, Ta, Sa, Sb, Vib). Widespread in Europe, also in Siberia and Algeria.

Biology: adults in broadleaf forests from May to August. A parasitoid of Cicadellidae Idiocerinae. Reared from *Rhytidodus decimusquartus*

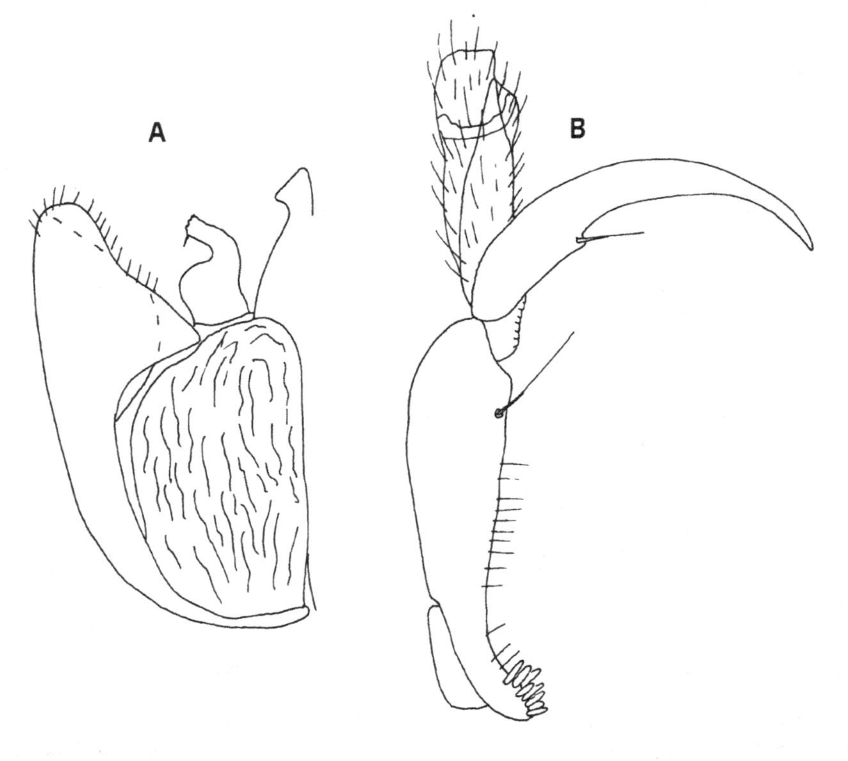

Fig. 35. *Anteon flavicorne* (Dalman): male genitalia (right half removed) (A) and chela (B).

(Schrank); *Idiocerus stigmaticalis* Lewis; *Populicerus laminatus* (Flor), *populi* (L.), *albicans* (Kirschbaum) and *confusus* (Flor); *Tremulicerus distinguendus* (Kirschbaum). The prepupa overwinters in a cocoon (Arzone, Alma and Arnò, 1987). This species may be parasitized by *Ismarus flavicornis* Thomson (Hymenoptera Diapriidae) (Chambers, 1955; Nixon, 1957).

13. *Anteon ephippiger* (Dalman, 1818)

Plates 12, 16, Fig. 36.

Gonatopus ephippiger Dalman, 1818: 81.
Gonatopus collaris Dalman, 1818: 82.
Dryinus facialis Thomson, 1860: 177.
Chelogynus rufovariegatus Berland, 1928: 164.

Female (Plate 12): fully winged; length 1.81-2.87 mm; colour very variable; head usually yellow or reddish (light or dark); occasionally head brown or black; antennae yellow or testaceous; occasionally antennal segments 1-2 or 3-10 brownish; pronotum whitish or yellow; occasionally pronotum reddish, or dark reddish, with brown spots; propectus yellow or reddish, occasionally blackish; scutum, scutellum and metanotum yellow or reddish, occasionally brown or black; mesopleura and metapleura black or reddish; propodeum black; gaster brown or black, occasionally reddish; legs yellow; tegulae testaceous; antennae long, with segment 1 as long as or slightly longer than segment 4 (8:6); head shining, smooth, unhaired, without sculpture, or weakly alutaceous, or very weakly granulated; pronotum with posterior surface not transverse, approximately as long as broad; pronotum, scutum, scutellum and metanotum shining, smooth, without sculpture; notauli incomplete, reaching approximately 0.3-0.5 length of scutum; propodeum reticulate rugose, with a strong transverse keel between dorsal and posterior surfaces; posterior surface with two longitudinal keels; median area usually rugose, occasionally shining and smooth; fore wing hyaline, without dark transverse bands; distal part of stigmal vein much shorter than proximal part; segment 1 of fore tarsus slightly shorter than segment 4; fore tarsal segment 2 produced into a hook; enlarged claw (Fig. 36 B) with a proximal prominence bearing a long bristle; segment 5 of fore tarsus (Fig. 36 B) with basal part

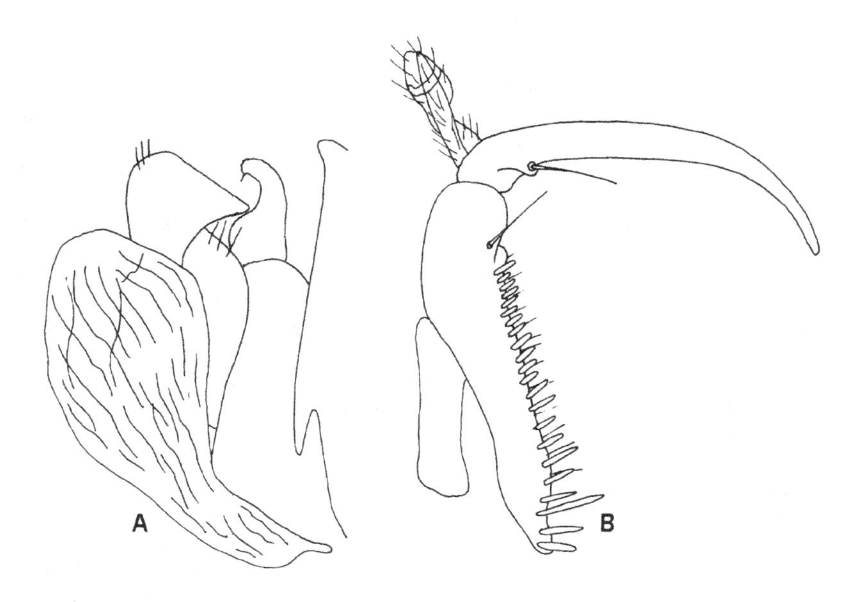

Fig. 36. *Anteon ephippiger* (Dalman): male genitalia (right half removed) (A) and chela (B).

shorter than distal part, with 1 row of 16-19 lamellae; distal apex with a group of 3 lamellae.

Male (Plate 16): fully winged; length 1.75-2.50 mm; black; antennae brown; legs testaceous, at most with hind coxae blackish proximally; tegulae testaceous; mandibles yellow, with teeth brown; head shining, smooth, very finely punctate, usually without frontal line; frons without lateral keels; scutum shining, smooth, with anterior half strongly punctate and with posterior half without punctures; notauli incomplete, reaching approximately mid-length of scutum; scutellum and metanotum shining, smooth, without sculpture; propodeum reticulate rugose, with a strong transverse keel between dorsal and posterior surfaces; posterior surface with two longitudinal keels; median area as rugose as lateral regions; fore wing hyaline, without dark transverse bands; distal part of stigmal vein much shorter than proximal part; parameres (Fig. 36 A) with a distal inner process pointed and extended apically; inner margin of this process convex or straight; dorsal proximal membranous process present.

Distribution: common in Denmark (SJ, EJ, F, NEZ), Sweden (Sk., Öl., Gtl., Vg., Nrk., Upl., Vb.), Norway (Bø, HOy) and East Fennoscandia (Ab, N, St, Ta, Sa, Tb, Om, Vib). In Norway probably also common in other provinces, in addition to Buskerud and Hordaland. Widespread in Europe, also in Korea, Mongolia, Siberia, Turkey, Lebanon, Morocco.

Biology: adults in broadleaf forests, pastures and fields from June to August. A parasitoid of Cicadellidae Deltocephalinae and Macropsinae. Reared from *Macrosteles sexnotatus* (Fallén) and *laevis* (Ribaut); *Opsius stactogalus* Fieber; *Macropsis* sp. Mating observed by Becker (1975, in Waloff and Jervis, 1987).

14. ***Anteon pubicorne*** (Dalman, 1818)

Plate 13, Fig. 37.

Gonatopus pubicornis Dalman, 1818: 87.
Dryinus tenuicornis Dalman, 1823: 13.
Gonatopus cephalotes Ljungh, 1824: 267.
Dryinus lucidus Haliday in Curtis, 1828: 206.

Female (Plate 13): fully winged; length 2.00-2.81 mm; black; occasionally head and thorax uniformly brown or with cupreous tinges; antennae testaceous, with segments 1-2 or only 1 yellow; occasionally antennae testaceous, with segments 3-6 brown; legs completely testaceous, occasionally with clubs of femora brown; tegulae testaceous; mandibles yellow, with brown teeth; antennae long, with segment 1 slightly longer than segment 4; proportions of antennal segments variable; head shining, smooth, finely or strongly punctate, without sculpture among punctures; frons usually with a short frontal line; this sometimes absent in small specimens; frons with two lateral keels near orbits; pronotum shining, strongly punctate; scutum shining, smooth, finely punctate, without sculpture among punctures; notauli incomplete, reaching approximately 0.5-0.6 length of scutum; scutellum and metanotum shining, smooth, finely punctate; propodeum reticulate rugose, with a strong transverse keel between dorsal and posterior surfaces; median area usually as rugose as lateral regions; occasionally smooth and shining, not rugose; fore wing hyaline, without dark transverse bands; distal part of stigmal vein much shorter than proximal part; fore tarsal segment 2 produced into a hook; segment 1 of fore tarsus approximately 0.66 as long as segment 4; enlarged claw (Fig. 37 B) with a proximal prominence bearing a long bristle; segment 5 of fore tarsus (Fig. 37 B) with basal part shorter than distal part, with 1 row of 20-23 lamellae; distal apex with a group of 5-6 lamellae.

Male: fully winged; length 1.37-2.87 mm; black; antennae brown or black; legs brown or black, with light tarsi and joints; tegulae testaceous; mandibles yellow, with brown teeth; proportions of antennal segments variable; head shining, with variable sculpture; in small specimens head smooth, finely punctate and without sculpture; in large specimens head smooth, but strongly punctate and without sculpture among punctures; scutum haired, shining, smooth, finely punctate; notauli incomplete, reaching approximately mid-length of scutum; scutellum and metanotum shining, smooth, finely punctate; propodeum reticulate rugose, with a strong transverse keel between dorsal and posterior surfaces; posterior surface with two longitudinal keels, these rarely invisible; median area rugose, dull or shining; fore wing hyaline, without dark transverse bands; distal part of stigmal vein much shorter than proximal part; parameres (Fig. 37 A) with a distal inner process pointed and extended apically; inner margin of this process convex or straight; dorsal proximal membranous process present.

Distribution: common in Denmark (not found only in WJ and NWZ), Sweden (Sk., Bl., Sm., Öl., Gtl., Ög., Vg., Boh., Sdm., Upl., Vrm., Gstr., Hls., Med., Jmt., Vb., Nb., Lu. Lpm.), Norway (Ø, AK, On, Bø, Bv, VE, AAy, HOy, Nsi) and East Fennoscandia (Al, Ab, N, St, Ta, Sa, Tb, Sb, Om, Ok, ObN, Li, Le, Vib, Kr, Lr). Widespread in Europe,

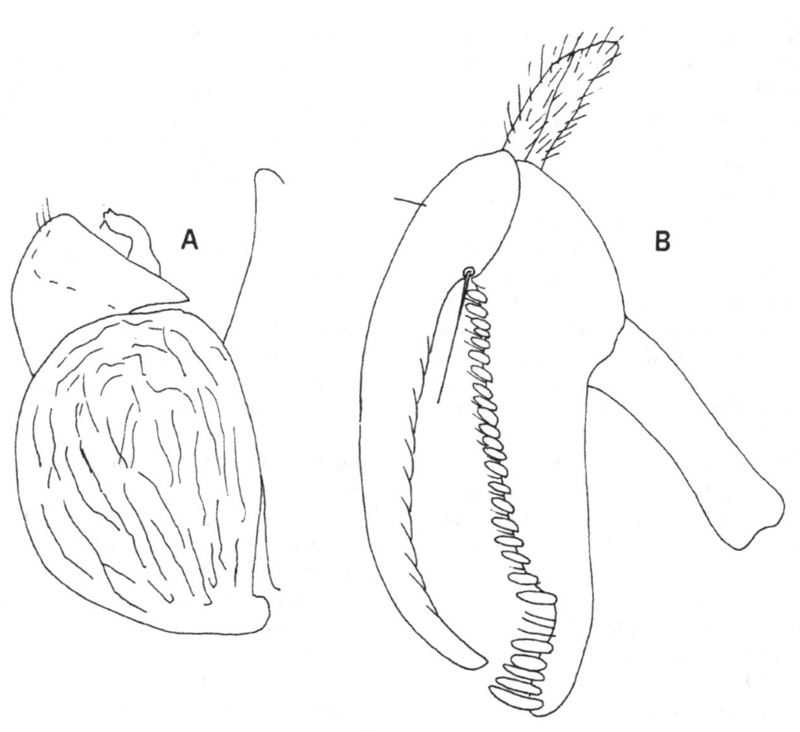

Fig. 37. *Anteon pubicorne* (Dalman): male genitalia (right half removed) (A) and chela (B).

also in Korea, Mongolia, Siberia, Armenia, Morocco.

Biology: adults in broadleaf forests, pastures and fields from May to September. A parasitoid of Cicadellidae Macropsinae and Deltocephalinae. Reared from *Macropsis* sp.; *Macrosteles sexnotatus* (Fallén) and *viridigriseus* (Edwards); *Psammotettix confinis* (Dahlbom) and *nodosus* (Ribaut); *Arocephalus punctum* (Flor); *Streptanus sordidus* (Zetterstedt); *Arthaldeus pascuellus* (Fallén); *Euscelis incisus* (Kirschbaum); *Opsius stactogalus* Fieber. This species overwinters in a cocoon spun on the host plant. The number of generations in Fennoscandia and Denmark is not known, but in England the species may be univoltine or bivoltine (Waloff and Jervis, 1987). It may be parasitized by *Ismarus rugulosus* Förster (Hymenoptera Diapriidae) (Waloff, 1975, misid. *Chelogynus lucidus* (Haliday)). Information on biology in Waloff (1974, 1975).

15. ***Anteon infectum*** (Haliday, 1837)

Fig. 38.

Dryinus infectus Haliday in Walker, 1837: 419.
Dryinus lateralis Thomson, 1860: 178.

Female: fully winged; length 3.00-4.50; black; antennae testaceous, with segments 6-10 or 7-10 fuscous dorsally; legs testaceous, with coxae partly brown, clubs of hind femora brown, occasionally also clubs of fore femora brown; tegulae testaceous; mandibles yellow, with brown teeth; antennae short, with segment 1 approximately twice as long as segment 4; antennal segments 2+3 approximately as long as, or slightly longer than, segment 1; head shining, smooth, punctate, with frons reticulate rugose; frons with a short median line and with two lateral keels; pronotum rugose, dull, smooth only near posterior margin; scutum shining, smooth, punctate; notauli incomplete, reaching approximately mid-length of scutum; scutellum and metanotum shining, smooth, punctate; propodeum reticulate rugose, with a strong transverse keel between dorsal and posterior surfaces; posterior surface with two longitudinal keels; median area shining, smooth, rugose only near margins; fore wing with a dark transverse band near pterostigma; distal part of stigmal vein much shorter than proximal part; fore tarsal segment 1 not quite as long as segment 4; fore tarsal segment 2 produced into a hook; enlarged claw (Fig. 38 B) with a proximal prominence bearing a long bristle; segment 5 of fore tarsus (Fig. 38 B) with basal part shorter than distal part, with 2 rows of 12-36 lamellae extending continuously to distal apex.

Male: fully winged; length 2.30-3.43 mm; black; antennae testaceous, with segments 1-8 or 1-10 fuscous dorsally; legs testaceous, with coxae partly brown or black and femoral clubs brown; occasionally clubs of fore femora testaceous and hind tibiae brown; tegulae testaceous; mandibles yellow, with brown teeth; head shining, punctate, with numerous short keels; frons with a short median line and with two lateral keels; pronotum dull, rugose; scutum shining, smooth, with anterior half strongly punctate and with posterior half finely punctate; notauli incomplete, reaching approximately mid-length of scutum; scutellum and metanotum finely punctate; propodeum reticulate rugose, with a strong transverse keel between dorsal and posterior surfaces; posterior surface with two longitudinal keels; median area shining, smooth, rugose only near margins; fore wing hyaline, without dark transverse bands; distal part of stigmal vein much shorter than proximal part; parameres (Fig. 38 A) with no distal inner process; dorsal proximal membranous process present.

Distribution: uncommon in Denmark (SJ, NEJ, F, LFM, NWZ), Sweden (Sk., Boh.), Norway (only found in Ø: Telemarkslunden, 19.V-17.VI.1992, by Lars Ove Hansen) and East Fennoscandia (only found in N: Helsinki). Widespread in Europe, also in Japan and Siberia.

Biology: adults in broadleaf forests from May to June. A parasitoid of Cicadellidae Iassinae. Reared from *Iassus lanio* (L.). According to Waloff and Jervis (1987) the species is probably truly monophagous. It may be parasitized by *Ismarus halidayi* Förster (Hymenoptera Diapriidae) (Chambers, 1981).

16. ***Anteon exiguum*** (Haupt, 1941)

Fig. 39.

Anteon subarcticus Hellén, 1935: 7 (nomen nudum).
Chelogynus exiguus Haupt, 1941: 52.
Anteon flaviscapus Jansson, 1950: 221.
Anteon subarcticus Hellén, 1953: 96.

Female: fully winged; length 1.31-2.31 mm; head black, with mandibles testaceous; antennae black, with segments 1-2 testaceous; thorax, propodeum and gaster black; legs testaceous; antennae thickened distally; proportions of antennal segments

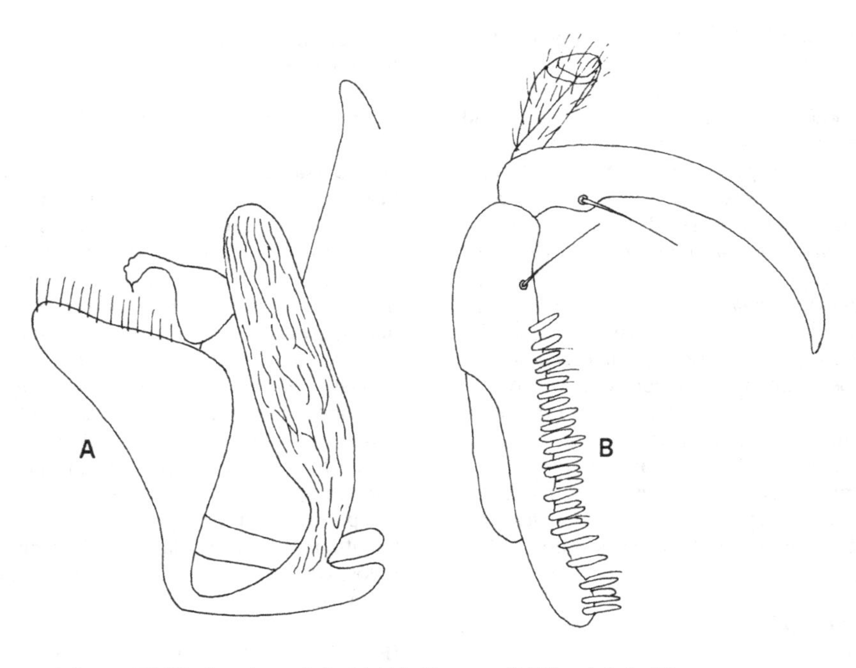

Fig. 38. *Anteon infectum* (Haliday): male genitalia (right half removed) (A) and chela (B).

variable; head shining, smooth, punctate, without sculpture among punctures; frontal line complete; occipital carina complete; POL = 5; OL = 3; OOL = 5; OPL = 3; TL = 3; pronotum with anterior surface rugose and dull; posterior surface transverse, much broader than long (Fig. 39 C), shining, smooth, without sculpture; notauli incomplete, reaching approximately 0.5 length of scutum; scutellum and metanotum shining, smooth, without sculpture; fore wing hyaline, without dark transverse bands; distal part of stigmal vein much shorter than proximal part; fore tarsal segments in following proportions: 5:2:2:5:11; segment 2 of fore tarsus produced into a hook; enlarged claw (Fig. 39 B) with a proximal prominence bearing a long bristle; segment 5 of fore tarsus (Fig. 39 B) with basal part shorter than distal part, with 1 row of approximately 11 lamellae; distal apex with a group of 2 lamellae.

Male: fully winged; length 1.62-2.18 mm; black; mandibles testaceous-reddish; antennae brown, with segment 1 testaceous; legs testaceous-reddish, with stalk of hind femora dark; antennae not thickened distally; antennal segments in following proportions: 7:5:6:6:6:6:6:5.5:8; head shining, finely punctate, with region behind ocelli and temples weakly rugose and granulated; occasionally head alutaceous; frons with a weak longitudinal furrow in front of anterior ocellus; an incomplete frontal line visible in this furrow; occipital carina complete; POL = 5, OL = 3; OOL = 3; OPL = 3; TL = 3; scutum, scutellum and metanotum shining, finely punctate, without sculpture among punctures; notauli incomplete, reaching approximately 0.5 length of scutum; propodeum reticulate rugose, with a strong transverse keel between dorsal and posterior surfaces; posterior surface with two longitudinal keels; median area as rugose as lateral regions; fore wing hyaline, without dark transverse bands; distal part of stigmal vein much shorter than proximal part; genital armature (Fig. 39 A) with a dorsal proximal membranous process reaching approximately 0.5 length of parameres; parameres with a distal inner process pointed and extended apically; inner margin of this process convex or straight.

Distribution: uncommon in Denmark (only found in LFM: Bøtø, 6.VIII), Sweden (Sk., Bl., Hall., Ög., Vg., Nrk., Sdm., Upl., Vstm., Vrm., Dlr., Hls.), Norway (Ø, AK, Bø, Bv, VE, TEi, AAy) and East Fennoscandia (only found in Ab, N, Li). Widespread in Central and Northern Europe, not found in Southern Europe; also in Korea. This species seems more uncommon in Central than in Northern Europe.

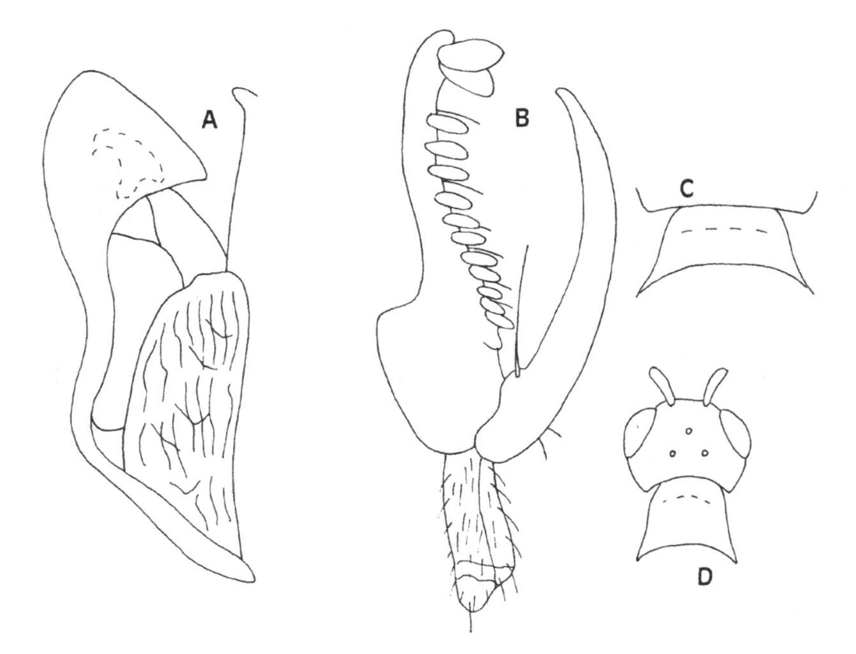

Fig. 39. *Anteon exiguum* (Haupt): male genitalia (right half removed) (A), chela (B) and female pronotum (in dorsal view) (C); D: head and pronotum (in dorsal view) of female of *Anteon pubicorne* (Dalman).

Biology: adults in broadleaf forests from May to September. Hosts unknown.

17. ***Anteon tripartitum*** Kieffer, 1905

Figs 2, 40.

Anteon tripartitus Kieffer in Kieffer & Marshall, 1905: 194.
Antaeon kiefferi Chitty, 1908: 143.

Female: fully winged; length 3.12-3.50 mm; black; antennae brown, with segments 1-2 or only 1 yellow; legs testaceous, with partly brown hind coxae and hind femoral clubs; tegulae testaceous; mandibles yellow, with brown teeth; antennae short, with segment 1 approximately twice as long as segment 4; head shining, smooth, granulated, with frons and occiput strongly punctate and with numerous irregular keels; vertex not punctate; ocellar triangle with keels joining ocelli; these keels occasionally weakly visible, but the keel joining posterior ocelli always visible; frontal line present; frons with two lateral keels; pronotum dull, rugose, with transverse keels, smooth near posterior margin; posterior surface of pronotum not transverse, approximately as long as broad; scutum shining, smooth, with posterior half strongly punctate and without sculpture among punctures; notauli incomplete, reaching approximately 0.5 length of scutum; scutellum and metanotum shining, smooth, finely punctate; propodeum reticulate rugose, with a strong transverse keel between dorsal and posterior surfaces; posterior surface with two longitudinal keels; median area as rugose as lateral areas; fore wing hyaline, without dark transverse bands; occasionally fore wing darkened distally; distal part of stigmal vein much shorter than proximal part; segment 2 of fore tarsus produced into a hook; segment 4 of fore tarsus slightly shorter than segment 1; enlarged claw (Fig. 40 B) with a proximal prominence bearing a long bristle; segment 5 of fore tarsus (Fig. 40 B) with basal part shorter than distal part, with 2 rows of 27-30 lamellae; distal apex with a group of 7-12 lamellae.

Male (Fig. 2): fully winged; length 2.37-3.00 mm; black; antennae black or brown; legs brown or black, with joints, tarsi and fore tibiae light; rarely legs completely testaceous, with hind coxae basally darkened; tegulae testaceous; mandibles yellow, with brown teeth; head dull, weakly granulated or alutaceous, with numerous irregular keels, punctures more numerous on frons; ocellar triangle with keels joining ocelli; frons with a median line and two lateral keels directed towards antennal sockets; scutum shining, smooth, haired, with anterior half strongly punctate and posterior half finely punctate or without punctures; notauli incomplete, reaching approximately 0.3-0.5 length of scutum; scutellum and metanotum shining, smooth, finely punctate; propodeum reticulate rugose, with a strong transverse keel between dorsal and posterior surfaces; posterior surface with two longitudinal keels; median area as rugose as lateral areas; fore wing hyaline, without dark transverse bands; distal part of stigmal vein much shorter than proximal part; parameres (Fig. 40 A) with inner distal process pointed and extended apically; inner margin of this process convex or straight; dorsal proximal membranous process present.

Distribution: uncommon in Denmark (SJ, F, LFM), Sweden (Ög., Nrk., Upl., Vstm.), Norway (Ø, Bø, AAy) and East Fennoscandia (Al, Ab, N, St, Sa, Vib, Kr). Widespread in Europe, also in Siberia.

Biology: adults in broadleaf forests and fields from May to July. A parasitoid of Cicadellidae Deltocephalinae. Reared from *Graphocraerus ventralis* (Fallén); *Thamnotettix confinis* (Zetterstedt) (Burn, 1993).

18. ***Anteon gaullei*** Kieffer, 1905

Plates 14, 17, Figs 31, 41.

Anteon gaullei Kieffer in Kieffer & Marshall, 1905: 161.

Anteon cameroni Kieffer in Kieffer & Marshall, 1905: 162.

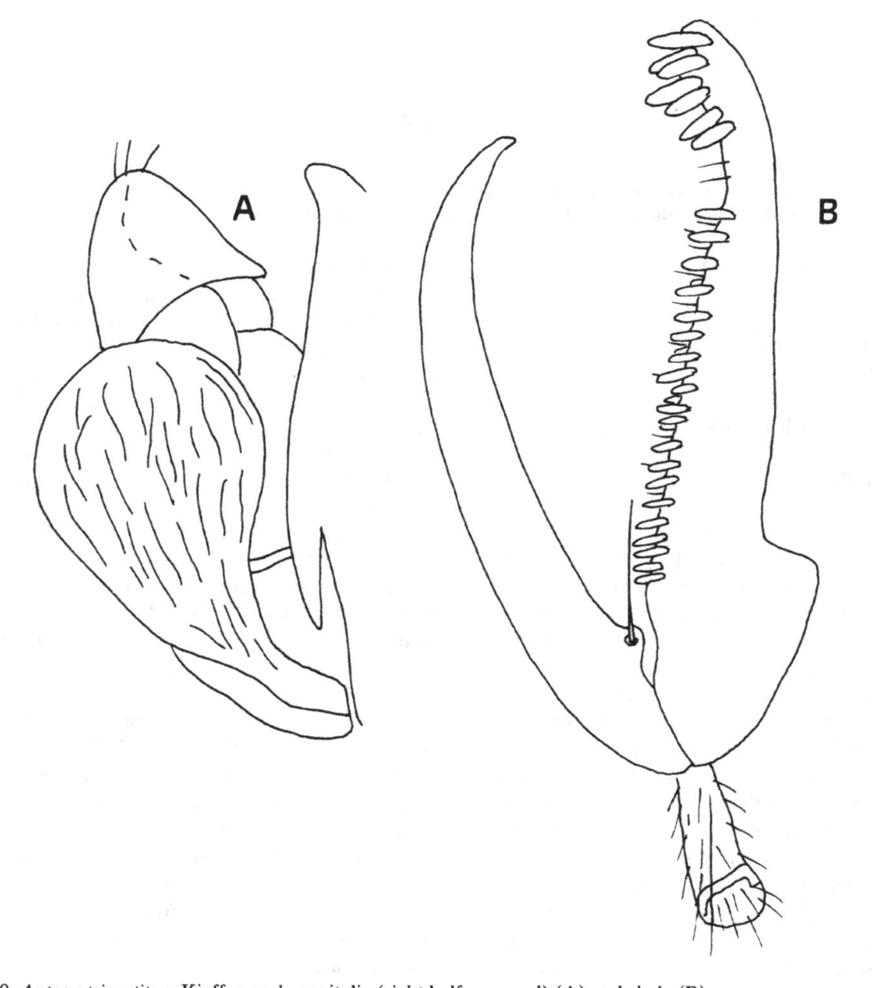

Fig. 40. *Anteon tripartitum* Kieffer: male genitalia (right half removed) (A) and chela (B).

Anteon maculipennis Kieffer in Kieffer & Marshall, 1905: 164.

Female (Plate 14, Fig. 31): fully winged; fore wings occasionally slightly shortened; length 2.37-4.00 mm; black; antennae brown, with segments 1-2 or 1-3 testaceous; mandibles yellow, with brown teeth; propectus yellow or black; pronotum yellow or reddish, rarely brown or dark-brown; scutum black, occasionally dark reddish; legs completely testaceous, occasionally with stalk of hind femora brown; gaster brown or black; tegulae testaceous; length of antennae variable; head shining, smooth, strongly punctate on frons, less strongly punctate on vertex and occiput, without sculpture among punctures; frons with a short median line, without lateral keels; pronotum shining, with anterior half strongly punctate and with posterior half finely punctate or without punctures; posterior surface of pronotum not transverse, approximately as long as broad; scutum, scutellum and metanotum shining, smooth, finely punctate, without sculpture among punctures; notauli incomplete, reaching approximately 0.5 length of scutum; propodeum reticulate rugose, with a strong transverse keel between dorsal and posterior surfaces; posterior surface with two longitudinal keels; median area as rugose as lateral regions; fore wing hyaline, without dark transverse bands; occasionally fore wing infuscate apically; distal part of stigmal vein much shorter than proximal part; proportions of fore tarsal segments variable: segment 1 of fore tarsus as long as, or shorter than, or longer than, segment 4; fore tarsal segment 2 produced into a hook; enlarged claw (Fig. 41 B) with a proximal prominence bearing a long bristle; segment 5 of fore tarsus (Fig. 41 B) with basal part shorter than distal part, with 2 rows of approximately 18-30 lamellae; distal apex with a group of 7-20 lamellae.

Male (Plate 17): fully winged; length 2.12-2.50 mm; black; antennae brown or light brown; mandibles yellow, with brown teeth; legs testaceous, with hind coxae brown proximally and occasionally with stalk of hind femora brown; tegulae testaceous; rarely body completely reddish; head shining,

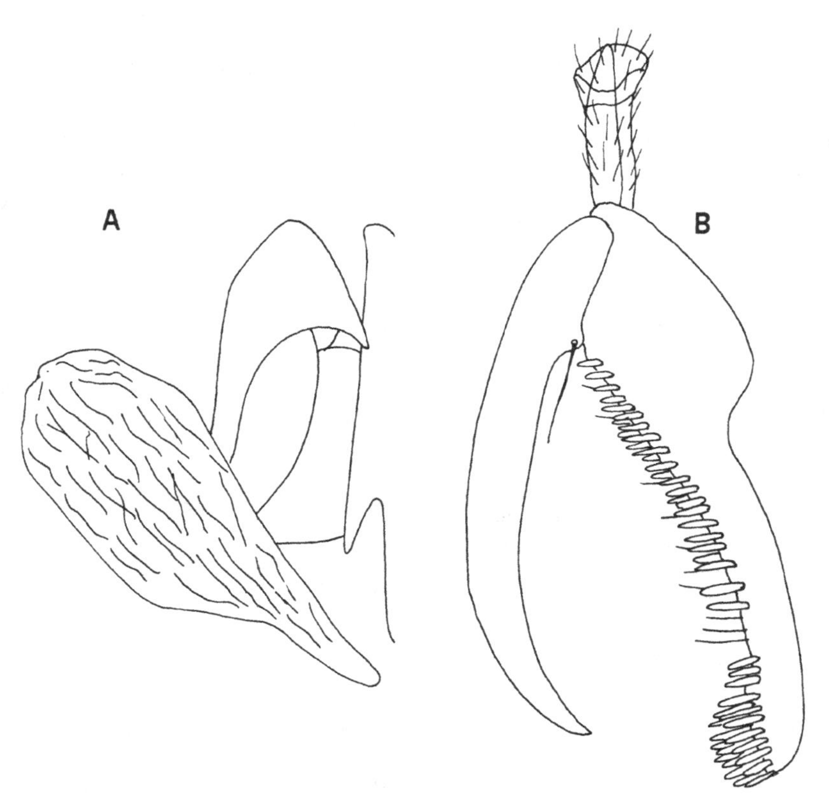

Fig. 41. *Anteon gaullei* Kieffer: male genitalia (right half removed) (A) and chela (B).

strongly punctate, without sculpture among punctures; scutum shining, smooth, with anterior half strongly punctate and with posterior half finely punctate or without sculpture; notauli incomplete, reaching approximately 0.5 length of scutum; scutellum and metanotum shining, smooth, without sculpture; propodeum reticulate rugose, with a strong transverse keel between dorsal and posterior surfaces; posterior surface with two longitudinal keels; median area as rugose as lateral regions; fore wing hyaline, without dark transverse bands; distal part of stigmal vein much shorter than proximal part; parameres (Fig. 41 A) with a distal inner process pointed and extended apically; inner margin of this process convex or straight; proximal dorsal membranous process present.

Distribution: common in Denmark (EJ, NWJ, F, LFM, NEZ), Sweden (Sk., Bl., Hall., Sm., Öl., Gtl., Ög., Vg., Nrk., Sdm., Vstm., Vrm.), Norway (Ø, AK, Bø, Bv, VE, AAy) and East Fennoscandia (Al, Ab, N, Ta, Sa, Oa, Tb, Vib). Widespread in Europe, also in Siberia.

Biology: adults in broadleaf forests from May to August. A parasitoid of Cicadellidae Macropsinae. Reared from *Macropsis* sp.

19. ***Anteon fulviventre*** (Haliday, 1828)

Plate 15, Fig. 19.

Dryinus fulviventris Haliday in Curtis, 1828: 206.
Dryinus fuscipes Thomson, 1860: 177.

Female (Plate 15): fully winged; length 2.00-3.00 mm; colour very variable; head yellow or reddish; occasionally head black or brown; antennae completely yellow or brown, with segments 6-10 or 7-10 yellow; pronotum and scutum yellow or light reddish; occasionally pronotum and scutum brown or black; scutellum and metanotum yellow or reddish; occasionally black or blackish; mesopleura and metapleura black or yellow; propodeum black, occasionally with yellow sides; gaster brown, or black, or reddish; legs testaceous; tegulae testaceous; mandibles yellow, with brown teeth; antennae short, with segment 1 approximately twice as long as segment 4; head dull, granulated, not punctate, smooth; frontal line short; pronotum granulated, dull; scutum shining, smooth, without sculpture or weakly alutaceous; notauli incomplete, reaching approximately 0.5 length of scutum; scutellum and metanotum shining, smooth, without sculpture; propodeum reticulate rugose, with a strong transverse keel between dorsal and posterior surfaces; posterior surface with two longitudinal keels; median area as rugose as lateral regions; fore wing hyaline, without dark transverse bands; distal part of stigmal vein much shorter than proximal part; segment 1 of fore tarsus slightly shorter than segment 4; fore tarsal segment 2 produced into a hook; enlarged claw (Fig. 19) with a proximal prominence bearing a long bristle; segment 5 of fore tarsus (Fig. 19) with basal part shorter than distal part, with 1 row of 20-21 lamellae; distal apex with a group of 3-6 lamellae.

Male: fully winged; length 1.87-2.50 mm; black; antennae brown-yellow; mandibles yellow, with brown teeth; legs completely yellow, occasionally with stalks of hind femora brown; tegulae testaceous; head dull, smooth, granulated, without punctures; frontal line present; frons without lateral keels; scutum shining, smooth, haired, without sculpture or weakly granulated; notauli incomplete, reaching approximately 0.5 length of scutum; scutellum and metanotum shining, smooth, without sculpture; propodeum reticulate rugose, with a strong transverse keel between dorsal and posterior surfaces; posterior surface with two longitudinal keels; median area as rugose as lateral regions; fore wing hyaline, without dark transverse bands; distal part of stigmal vein much shorter than proximal part; parameres (Fig. 19) with distal inner process pointed and extended apically; inner margin of this process convex or straight; proximal dorsal membranous process present, but very short.

Distribution: common in Denmark (SJ, EJ, F, LFM, NEZ), Sweden (Sk., Bl., Hall., Sm., Ög., Vg., Nrk., Upl., Vstm., Gstr.), Norway (Ø, AK, HEs, Bø, AAy) and East Fennoscandia (Al, Ab, N, St, Ta, Sa, Tb, Vib). Widespread in Europe, also in Algeria and Morocco.

Biology: adults in broadleaf forests, pastures and fields from May to September. A parasitoid of Cicadellidae Deltocephalinae. Reared from *Macrosteles frontalis* (Scott).

3. Subfamily Dryininae

Type genus: ***Dryinus*** Latreille, 1804.

Female: fully winged; fore wing with costal, basal and subbasal cells fully enclosed by pigmented veins; fore tarsus chelate; chela with rudimentary claw; mandibles with 1-4 teeth; pronotal tubercles present; tibial spurs usually 1, 1, 2, occasionally 1, 1, 1.

Male: fully winged; fore wing with costal, basal and subbasal cells fully enclosed by pigmented veins; mandibles with 3-4 teeth; tibial spurs 1, 1, 2.

Distribution: worldwide.

Hosts: Fulgoromorpha.

Genera: one genus in Fennoscandia and Denmark.

4. Genus *Dryinus* Latreille, 1804

Dryinus Latreille, 1804: 176.
Type-species: *Sphex collaris* Linnaeus, 1767: 946, first included species.
Campylonyx Westwood, 1835: 52.
Type-species: *Campylonyx ampuliciformis* Westwood, 1835: 52, by monotypy.
Mesodryinus Kieffer in Kieffer & Marshall, 1906: 497.
Type-species: *Dryinus niger* Kieffer, 1904: 352, by subsequent designation.
Lestodryinus Kieffer, 1911: 108 (new name for *Dryinus* Latreille).
Type-species: *Sphex collaris* Linnaeus, 1767: 946, first included species.

Female: enlarged claw as long as, or shorter than, fore tibia; antennae with no tufts of long hair; maxillary palps 6-segmented; labial palps 3-segmented.

Male: mandibles tridentate; maxillary palps 6-segmented; labial palps 3-segmented.

Distribution: worldwide.

Hosts: Fulgoromorpha.

Species: one species in Fennoscandia and Denmark.

20. *Dryinus niger* Kieffer, 1904
Plates 18-19, Fig. 42.

Dryinus niger Kieffer, 1904: 352.
Mesodryinus brittanicus Richards, 1939: 228.

Female (Plate 18): fully winged; length 4.00-4.37 mm; head black, with yellow clypeus and mandibles; antennae testaceous; thorax and propodeum black; gaster brown; legs testaceous, with brown femoral clubs and hind femora; distal part of median and hind tibiae brown; antennal segments in following proportions: 9:5:23:11:10:9:7:7:6:7; head granulated, dull, with numerous weak longitudinal striae on frons and vertex; occipital carina incomplete, only visible behind ocelli; posterior ocelli touching occipital carina; pronotum dull, haired, granulated, with a strong posterior transverse impression and with a weak anterior transverse impression; pronotal disc weakly humped; posterior collar of pronotum very short and reduced; pronotal tubercles not reaching tegulae; scutum dull, completely reticulate rugose; notauli absent; scutellum, metanotum and propodeum dull, reticulate rugose; fore wing with two brown transverse bands; distal part of stigmal vein slightly longer than proximal part; segment 1 of fore tarsus longer than segment 4 (14:11); segments 2 and 3 of fore tarsus produced into hooks; enlarged claw (Fig. 42 B) without a subapical tooth, with 6 lamellae; segment 5 of fore tarsus (Fig. 42 B) with 2 rows of 10-18 lamellae near distal apex; tibial spurs 1, 1, 2.

Male (Plate 19): fully winged; length 2.87-3.37 mm; black; clypeus and mandibles testaceous; antennae brown; legs yellow, with hind coxae, clubs of hind femora and distal half of hind tibiae brown; antennae not thickened distally, with segment 3 more than eight times as long as broad (22:2.5); antennal segments in following proportions: 7:5:22:13:15:13:11:10:9:12; head dull, granulated and reticulate rugose; frontal line absent; occipital carina incomplete, completely visible on dorsal side of head, touching eyes laterally; occipital carina not visible on genae; temples absent; POL = 8; OL = 3; OOL = 5; posterior ocelli touching occipital carina; mandibles with 1- 2 teeth; scutum dull, granulated and reticulate rugose; notauli incomplete, weakly visible and reaching approximately 0.4-0.5 length of scutum; scutellum and metanotum dull, granulated; metanotum with two unsculptured lateral areas; propodeum dull, reticulate rugose, without transverse or longitudinal keels; fore wing hyaline, without dark transverse bands; distal part of stigmal vein slightly longer than proximal part (10:9); genitalia as in fig. 42 A; tibial spurs 1, 1, 2.

Distribution: uncommon in Denmark (only found in SJ: Sotrup), Norway (only found in TEi: Notodden, Lisleherad, 22.VI-6.VIII.1993, by A. Bakke)

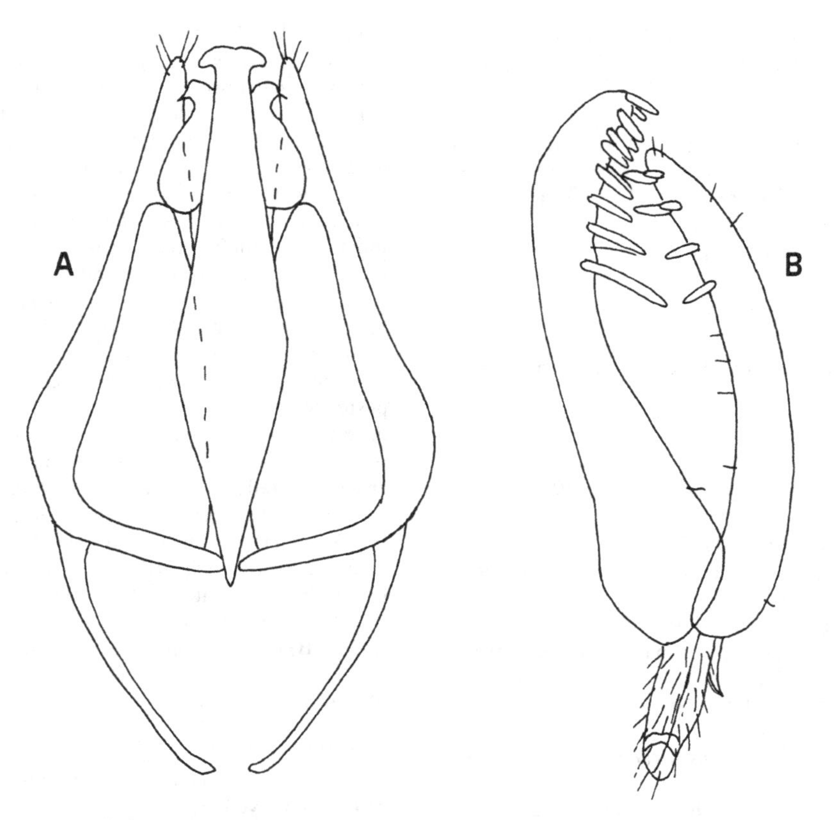

Fig. 42. *Dryinus niger* Kieffer: male genitalia (A) and chela (B).

and Sweden (only found in Sm., Vstm. (Solbacken), Vrm. (Ekshärad)). Not found in East Fennoscandia. France, England, Italy, Greece, Cyprus.

Biology: adults in fields and pastures from June to August. A parasitoid of Cixiidae.

4. Subfamily Gonatopodinae

Type genus: ***Gonatopus*** Ljungh, 1810

Female: usually apterous, occasionally fully winged, rarely micropterous; fore wing with costal, basal and subbasal cells fully enclosed by pigmented veins; maxillary palps composed of 2-6 segments; labial palps composed of 1-3 segments; mandibles quadridentate; fore tarsus chelate; chela with rudimentary claw; ocelli present; occipital carina absent or briefly visible behind and on sides of ocellar triangle; fore trochanters more than twice as long as broad; pronotal tubercles absent; in fully winged specimens metanotum very reduced and pterostigma very narrow; tibial spurs 1, 0, 1 or 1, 0, 2.

Male: fully winged; fore wing with costal, basal and subbasal cells completely enclosed by pigmented veins; maxillary palps composed of 2-6 segments; labial palps composed of 1-3 segments; mandibles tridentate; occipital carina absent or briefly visible behind and on sides of ocellar triangle; rarely occipital carina complete; pterostigma very narrow; occiput usually concave, occasionally straight; parameres with a dorsal process; tibial spurs 1, 1, 2.

Distribution: worldwide.

Hosts: Fulgoromorpha and Cicadomorpha (except for Cicadellidae Typhlocybinae, Idiocerinae, Macropsinae).

Genera: 2 genera in Fennoscandia and Denmark.

5. Genus *Haplogonatopus* Perkins, 1905

Haplogonatopus Perkins, 1905: 39.
Type-species: *Haplogonatopus apicalis* Perkins, 1905, by original designation.
Monogonatopus Richards, 1939: 200.
Type-species: *Gonatopus oratorius* Westwood, 1833: 496, by monotypy.

Female: apterous; maxillary palps 2-segmented; labial palps 1-segmented; tibial spurs 1, 0, 1.

Male: fully winged; maxillary palps 2-segmented; labial palps 1-segmented.

Distribution: worldwide, except for Nearctic region.

Hosts: Delphacidae.

Species: one species in Fennoscandia and Denmark.

21. *Haplogonatopus oratorius* (Westwood, 1833)

Plates 20-21, Figs 20, 43-44.

Gonatopus oratorius Westwood, 1833b: 496.
Gonatopus mayeti Kieffer in Kieffer & Marshall, 1905: 103.
Haplogonatopus atratus Esaki & Hashimoto, 1932: 25.

Female (Plate 20, Fig. 43): apterous; length 2.00-3.12 mm; head brown or black, with mandibles, clypeus and anterior region of frons testaceous; antennae brown, with segments 1-2 or 1-3 and 10 yellow; thorax and propodeum testaceous-reddish or testaceous; petiole and gaster black; legs testaceous-yellow; rarely body completely black; head excavated, shining, weakly granulated; pronotum not crossed by a transverse impression, shining, smooth, unhaired, without sculpture; scutum dull, unhaired, with a few longitudinal keels; metanotum, pleura and posterior surface of propodeum transversely striate, without sculpture among striae; anterior surface of metathorax + propodeum without sculpture; meso-metapleural suture obsolete; fore tarsal segments in following proportions: 13:2:3:8:13; enlarged claw (Fig. 20) with a subapical tooth and a row of 4-5 lamellae; segment 5 of fore tarsus (Fig. 20) with two rows of 6-12 lamellae; distal apex with a group of 3-9 lamellae.

Male (Plate 21, Fig. 44): fully winged; length 2.18-2.50 mm; black, with antennae brown; mandibles yellow-brown; legs fully yellow or brown, with light tarsi; antennal segment 3 more than four times as long as broad (4.4); head shining, haired, granulated; frontal line absent; temples distinct; occipital carina absent; scutum dull, granulated; notauli distinct, complete, strongly converging, but running parallel near posterior margin of scutum; scutellum and metanotum dull, haired, weakly punctate; propodeum reticulate rugose, without keels; fore wing hyaline, without dark transverse bands; dorsal process of parameres (Fig. 20) long, slender, with distal apex serrate.

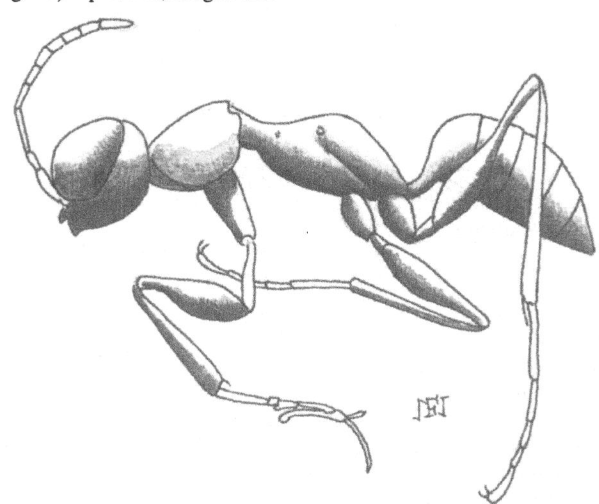

Fig. 43. Female of *Haplogonatopus oratorius* (Westwood) in lateral view.

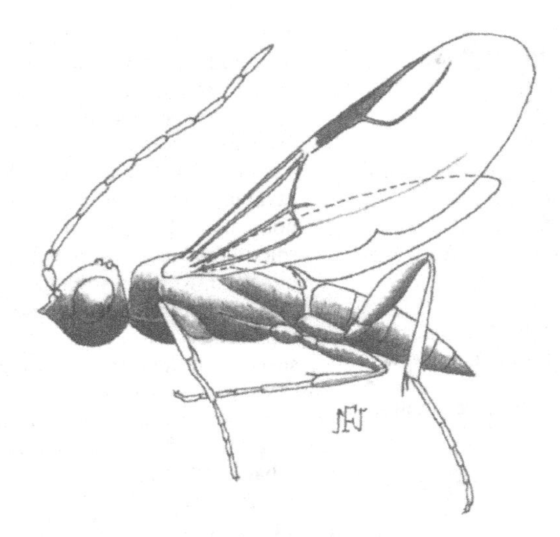

Fig. 44. Male of *Haplogonatopus oratorius* (Westwood) in lateral view.

Distribution: found only in East Fennoscandia (Vib). Widespread in Europe, also in the Mariana Islands, Japan, Korea, Siberia, Turkey, Lebanon, Canary Islands.

Biology: adults in pastures and fields from June to September. A parasitoid of Delphacidae. Reared from *Megadelphax sordidulus* (Stål); *Laodelphax striatellus* (Fallén); *Javesella pellucida* (F.). Information on biology by Kitamura (1982, 1983, 1985, 1986, 1989a, 1989b) (misid. *Haplogonatopus atratus* Esaki & Hashimoto).

6. Genus *Gonatopus* Ljungh, 1810

Gonatopus Ljungh, 1810: 161.
  Type-species: *Gonatopus formicarius* Ljungh, 1810, by monotypy.
Dicondylus Haliday in Curtis, 1829-30: 110.
  Type-species: *Dryinus bicolor* Haliday in Curtis, 1828: 206, by subsequent designation.
Labeo Haliday, 1833: 273.
  Type-species: *Labeo vitripennis* Haliday, 1833: 273, by monotypy.
Pseudogonatopus Perkins, 1905: 34.
  Type-species: *Pseudogonatopus kurandae* Perkins, 1905: 35, by original designation.
Neogonatopus Perkins, 1905: 42.
  Type-species: *Neogonatopus ombrodes* Perkins, 1905: 43, by original designation.
Platygonatopus Kieffer in Kieffer & Marshall, 1906: 500.
  Type-species: *Gonatopus planiceps* Kieffer, 1904: 355, by subsequent designation.
Donisthorpina Richards, 1939: 201.
  Type-species: *Donisthorpina formicicola* Richards, 1939: 201, by original designation.
Tetrodontochelys Richards, 1939: 217.
  Type-species: *Gonatopus ljunghii* Westwood, 1833: 496, by original designation.

Female: apterous; palpal formula 3/2, 4/2, 5/2, 4/3, 5/3, 6/3; enlarged claw with distal apex pointed; tibial spurs 1, 0, 1.

Male: fully winged; palpal formula 3/2, 4/2, 5/2, 4/3, 5/3, 6/3.

Distribution: worldwide .

Hosts: Fulgoromorpha.

Species: 14 species in Fennoscandia and Denmark.

Key to the species of *Gonatopus*

Females

1 Enlarged claw with a large subapical tooth (Figs 45 B, 47 B) 2
– Enlarged claw with a small subapical tooth (Figs 54 B, 56 B) 7

2 Pronotum not crossed by a strong transverse impression or very weakly impressed (Fig. 46) 3
– Pronotum crossed by a strong transverse impression (Fig. 10) 4
3 Metathorax + propodeum and scutellum reddish or yellow-reddish . . . . 22. *helleni* (Raatikainen)
– Metathorax + propodeum and scutellum black . . . . 23. *bicolor* (Haliday)
4 Maxillary palps 5-segmented 5
– Maxillary palps 2- or 3- or 4-segmented . . 6
5 Mesosoma testaceous and more or less brown (Plate 24); occasionally almost completely brown or black . . . . . 24. *pallidus* (Ceballos)
– Mesosoma completely testaceous or yellow (Plate 25) . . . . . 25. *formicicolus* (Richards)
6 Sides of metanotum prominent and often pointed (Plate 26) . . . . 26. *dromedarius* (Costa)
– Sides of metanotum rounded (Plate 27) . . . . 27. *distinctus* Kieffer
7 Pronotum not crossed by a strong transverse impression or weakly impressed (Fig. 53) . . . . 28. *pedestris* Dalman
– Pronotum crossed by a strong transverse impression (Fig. 55) 8
8 Segment 5 of fore tarsus with lamellae situated on a distinct prominence (Figs 58 C, 59 B) 9
– Segment 5 of fore tarsus with lamellae not situated on a distinct prominence (Figs 58 B, 60 B) 10
9 Colour mostly yellow-testaceous; only petiole and gaster black; occasionally propodeum with two brown lateral spots; anterior surface of metathorax + propodeum shining, without sculpture . . . . 31. *spectrum* (Van Vollenhoven)
– Colour very variable, but at least scutellum and metathorax + propodeum black; anterior surface of metathorax + propodeum usually dull, granulated, rarely shining and unsculptured . . . . 32. *distinguendus* Kieffer
10 Meso-metapleural suture obsolete . . . . 29. *lunatus* Klug
– Meso-metapleural suture distinct at least proximally 11
11 Segment 5 of fore tarsus serrate proximally (Fig. 57 C) . . . . 30. *striatus* Kieffer
– Segment 5 of fore tarsus not serrate proximally (Figs 58 B, 60 B) 12
12 Posterior surface of metathorax + propodeum not transversely striate . . . . 33. *formicarius* Ljungh
– Posterior surface of metathorax + propodeum strongly transversely striate . . . . 34. *clavipes* (Thunberg)

The female of *Gonatopus albifrons* Olmi is unknown.

Males

1 Antennae very slender, with segment 3 four or more than four times as long as broad (see plate 21) 2
– Antennae less slender, with segment 3 less than three and a half times as long as broad (Plate 36) 7
2 Dorsal process of parameres very reduced (Fig. 48 A) or very short (Fig. 49 A) 3
– Dorsal process of parameres very long (Figs 45 A, 47 A, 52 A, 54 A) 4
3 Dorsal process of parameres with distal apex rounded (Fig. 48 A) . . . . 24. *pallidus* (Ceballos)
– Dorsal process of parameres with distal apex pointed (Fig. 49 A) . . . . 25. *formicicolus* (Richards)
4 Dorsal process of parameres with apical and inner margins serrate (Fig. 52 A) . . . . 27. *distinctus* Kieffer
– Dorsal process of parameres with apical and inner margins not serrate (Figs 45 A, 47 A, 54 A) 5
5 Dorsal process of parameres slender and with distal apex pointed (Fig. 54 A) . . . . 28. *pedestris* Dalman
– Dorsal process of parameres less slender and with distal apex broadened and rounded (Figs 45 A, 47 A) 6
6 Minimum distance between notauli approximately as long as breadth of ocelli . . . . 23. *bicolor* (Haliday)
– Minimum distance between notauli shorter than breadth of ocelli . . . . 22. *helleni* (Raatikainen)
7 Notauli incomplete 8
– Notauli complete 9
8 Dorsal process of parameres transverse (Fig. 60 A); head without a prominent apophysis on sides of posterior ocelli (Fig. 56 D) . . . . 34. *clavipes* (Thunberg)
– Dorsal process of parameres long and slender, not transverse (Fig. 51 A); head with a very prominent apophysis on sides of posterior ocelli (see Fig. 56 C) . . . . 26. *dromedarius* (Costa)
9 Notauli more separated posteriorly; mini-

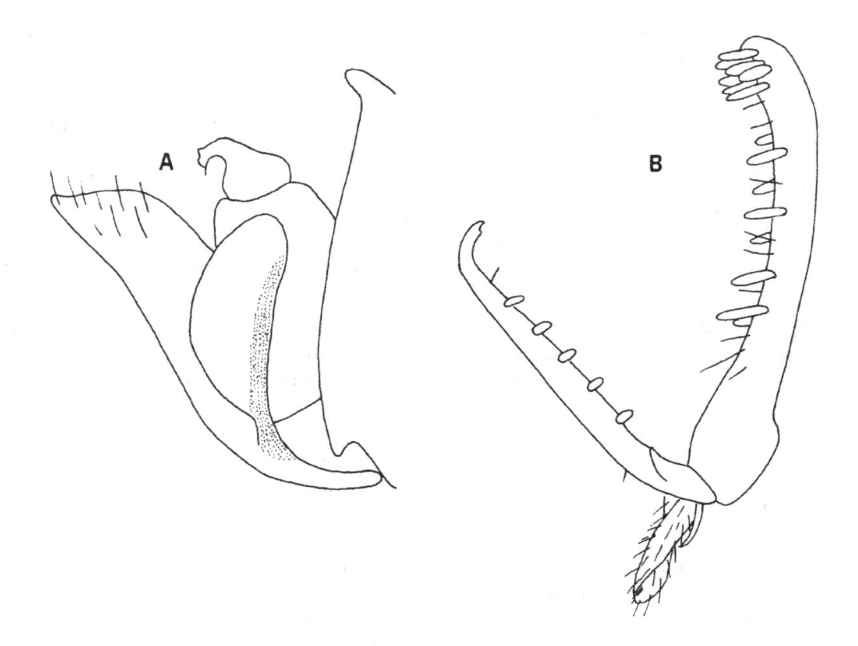

Fig. 45. *Gonatopus helleni* (Raatikainen): male genitalia (right half removed) (A) and chela (B).

mum distance between notauli approximately as long as antennal segment 2 .... 32. *distinguendus* Kieffer

– Notauli closer or joined posteriorly; minimum distance between notauli much shorter than antennal segment 2 10

10 Head with a very prominent apophysis on sides of posterior ocelli (Fig. 56 C) 11

– Head without a prominent apophysis on sides of posterior ocelli (Fig. 56 D) 13

11 Head black, with mandibles, clypeus and a U-shaped frontal dark yellow-whitish; dorsal process of parameres broadened (Fig. 60 C) .... 35. *albifrons* Olmi

– Head black except for testaceous mandibles; occasionally clypeus and malar space also testaceous; never with a U-shaped frontal mark yellow-whitish; dorsal process of parameres slender (Figs 51 A, 56 A) 12

12 Head black, with only mandibles testaceous .... 29. *lunatus* Klug

– Head black, with mandibles and clypeus testaceous; occasionally also malar space testaceous .... 26. *dromedarius* (Costa)

13 Dorsal process of parameres transverse (Fig. 60 A) .... 34. *clavipes* (Thunberg)

– Dorsal process of parameres not transverse (Figs 57 A, 58 A) 14

14 Dorsal process of parameres shorter (Fig. 58 A); areolae of propodeum small .... 33. *formicarius* Ljungh

– Dorsal process of parameres longer (Fig. 57 A); areolae of propodeum large .... 30. *striatus* Kieffer

The male of *Gonatopus spectrum* (Van Vollenhoven) is unknown.

22. *Gonatopus helleni* (Raatikainen, 1961)

Plate 22, Fig. 45.

Gonatopus dichromus Kieffer in Kieffer & Marshall, 1906: 505 (preoccupied).
Gonatopus rufescens Hellén, 1935: 8 (nomen nudum).
Dicondylus helleni Raatikainen, 1961: 131.

Female (Plate 22): apterous; length 2.50-3.25 mm; reddish or testaceous-yellow, with antennal segments 3-9 brown; occasionally vertex of head brown; petiole black; gaster reddish-yellow or brown-black, occasionally with first tergite reddish; head dull, excavated, granulated; pronotum dull, granulated, not crossed by a strong transverse impression, sometimes weakly impressed; scutellum dull, without sculpture; metathorax + propodeum dull, granulated, with posterior surface and pleura transversely striate; meso-metapleural suture ob-

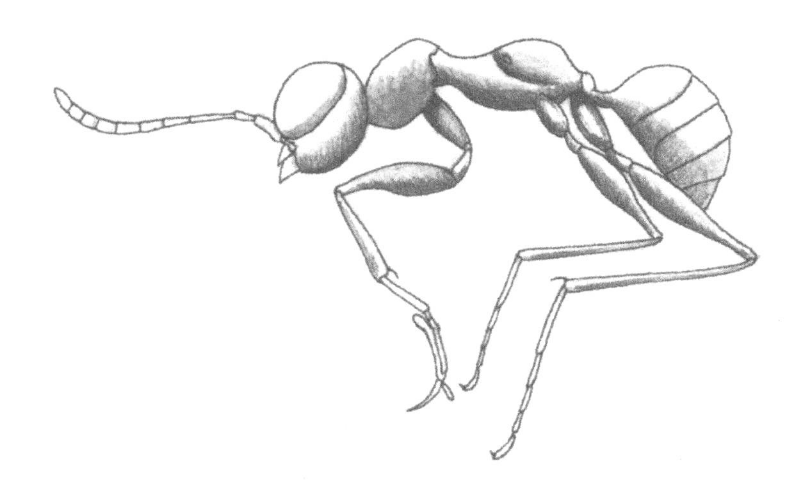

Fig. 46. Female of *Gonatopus bicolor* (Haliday) in lateral view.

solete; segment 1 of fore tarsus approximately twice as long as segment 4 (12:6); enlarged claw (Fig. 45 B) with a large subapical tooth and a row of 4-5 lamellae; segment 5 of fore tarsus (Fig. 45 B) with 2 rows of 6- 9 lamellae; distal apex with a group of 3-7 lamellae; maxillary palps with 2-4 segments; labial palps with 2 segments.

Male: fully winged; length 2.37-2.62 mm; black; mandibles yellow; antennae and tegulae brown; legs brown-yellow; antennal segment 3 more than four times as long as broad (5.25); head shining, haired, granulated; frontal line absent; temples distinct; occipital carina absent; occiput concave; scutum shining, granulated; notauli distinct, complete, separated posteriorly; minimum distance between notauli shorter than breadth of ocelli; scutellum and metanotum shining, smooth, punctate; propodeum reticulate rugose, without keels; fore wing hyaline, without dark transverse bands; dorsal process of parameres (Fig. 45 A) with paddle-shaped distal apex; maxillary palps 4-segmented; labial palps 2-segmented.

Distribution: rather common in Denmark (F, NEZ), Sweden (Sk., Bl., Sm., Öl., Ög., Vg., Nrk., Sdm., Upl., Vstm.), Norway (On, Bø) and East Fennoscandia (Ab, N, Ta, Sa, Oa, Tb). Widespread in Europe, also in Siberia.

Biology: adults in pastures and fields from June to September. A parasitoid of Delphacidae. Reared from *Megadelphax sordidulus* (Stål) and *Unkanodes excisa* (Melichar). This species may be preyed on by *Achorolophus gracilipes* (Kramer) (Acarina, Erythraeidae) (Raatikainen, 1961). The life cycle was described in Finland by Raatikainen (1961).

23. ***Gonatopus bicolor*** (Haliday, 1828)

Plate 23, Figs 46-47.

Dryinus bicolor Haliday in Curtis, 1828: 206.
Gonatopus conjunctus Kieffer in Kieffer & Marshall, 1905: 119.
Dicondylus lindbergi Heikinheimo, 1957: 78.

Female (Plate 23, Fig. 46): apterous; length 2.75-3.00 mm; head brown-black, with mandibles, clypeus and anterior part of frons (more along orbits) yellow; antennae brown, with segments 1-2 and occasionally 10 yellow or whitish; pronotum completely testaceous-yellow; occasionally pronotal disc yellow and sides black; rarely pronotum almost completely black; scutum yellow, with sides black; scutellum and metathorax + propodeum black, with distal apex of propodeum testaceous; gaster black; legs yellow or yellow-brown, occasionally with clubs of femora partly or completely brownish; head excavated, granulated; pronotum not crossed by a strong transverse impression or very weakly impressed, shining, granulated; scutum granulated; metanotum transversely striate; posterior surface of metathorax + propodeum and pleura transversely striate; meso-metapleural suture obsolete, rarely distinct, but very weak; fore tarsal segment 1 approximately twice as long as segment 4; enlarged claw (Fig. 47 B) with a large subapical tooth and a row of 4-5 lamellae; segment 5 of fore tarsus (Fig. 47 B) with two rows of 6-9

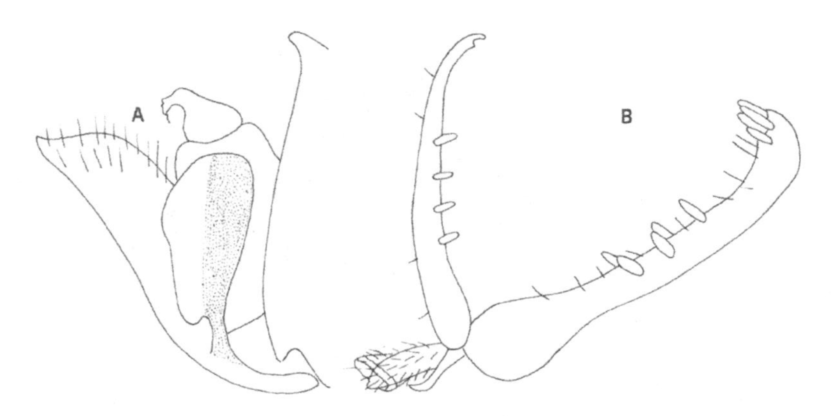

Fig. 47. *Gonatopus bicolor* (Haliday): male genitalia (right half removed) (A) and chela (B).

lamellae; distal apex with a group of 3-6 lamellae; maxillary palps with 2-4 segments; labial palps 2-segmented.

Male: fully winged; length 1.57-2.00 mm; head black, with mandibles yellow; antennae black-brown, with segments 1-2 partly yellow; thorax and propodeum black; gaster brown-black; legs yellow, with part of coxae and femora brown; antennae not thickened distally; antennal segment 3 four times as long as broad; antennal segments in following proportions: 5:4.5:7:8:7.5:7.5:6.5:6.5:6:8.5; head shining, granulated; frontal line absent; temples distinct; occipital carina absent; no shining oval area visible on sides of posterior ocelli; scutum shining; notauli complete, separated posteriorly; minimum distance between notauli as long as breadth of ocelli; scutellum and metanotum shining, weakly punctate; propodeum reticulate rugose; fore wing hyaline, without dark transverse bands; dorsal process of parameres (Fig. 47 A) with a paddle-shaped distal apex; maxillary palps 3-segmented; labial palps 2-segmented.

Distribution: rather common in Denmark (NWZ, NEZ), Sweden (Sk., Sm., Ög., Nrk., Sdm., Upl., Hls.), Norway (VE, Ry) and East Fennoscandia (Al, Oa, Vib). Widespread in Europe.

Biology: adults in pastures and fields from June to September. A parasitoid of Delphacidae. Reared from *Delphacinus mesomelas* (Boheman); *Ditropis pteridis* (Spinola); *Stiroma bicarinata* (H.- S.); *Unkanodes excisa* (Melichar); *Megadelphax sordidulus* (Stål); *Hyledelphax elegantulus* (Boheman); *Gravesteiniella boldi* (Scott); *Dicranotropis hamata* (Boheman); *Criomorphus albomarginatus* Curtis; *Javesella pellucida* (F.) and *obscurella* (Boheman); *Ribautodelphax collinus* (Boheman) and *pungens* (Ribaut). The life cycle was described by Raatikainen (1967), Lindberg (1950) and Heikinheimo (1957) in Finland, where this species is univoltine. In England *G. bicolor* is either bi- or trivoltine and hibernates as a first instar larva within the overwintering hosts (Waloff, 1974, 1975). Mating observed by Waloff (1974).

24. ***Gonatopus pallidus*** (Ceballos, 1927)

Plate 24, Fig. 48.

Dicondylus pallidus Ceballos, 1927: 107.

Female (Plate 24): apterous; length 2.18-3.31 mm; head testaceous, with a more or less large brown spot on vertex; antennae black, with segments 1-2 testaceous; mesosoma testaceous, with scutellum and posterior half of propodeum brown; occasionally metathorax + propodeum completely brown or black; occasionally pronotum with dark spots; gaster brown, with proximal half or third testaceous; legs yellow, with clubs of fore femora partly brown; antennae thickened distally; antennal segments in following proportions: 7:4.5:10:5.5:5:4.5:4:4:4:7; head excavated, shining, without sculpture, granulated except on occiput and anterior third of frons; POL = 1; OL = 2; OOL = 5; occipital carina absent; frontal line incomplete, not visible on anterior half of frons; temples distinct; pronotum crossed by a strong transverse impression, shining, without sculpture or weakly granulated; scutellum inclined; meso-metapleural suture obsolete; metathorax + propodeum shining, with anterior surface unsculptured or very weakly granulated; pleura and posterior surface of propodeum transversely striate; fore tarsal segments in following proportions: 12.5:2:4:10:15; enlarged claw (Fig. 48 B) with a large subapical tooth and a row of 4-5 lamellae; segment 5 of fore tarsus (Fig. 48 B) with 2 rows of 17-20 lamellae; distal apex with a group of approximately 6-15 lamellae; maxillary palps 5-segmented; labial palps 2-segmented.

Male: fully winged; length 2.18-2.43 mm; black; mandibles testaceous; legs testaceous, with partly brown coxae and tibiae; antennae not thickened distally, with segment 3 more than four times as long as broad (4.1); antennal segments in following proportions: 5:4:7.5:8:8:8:7.5:7.5:7:9; head shining, without sculpture; temples distinct; occipital carina absent; frontal line absent; frons with a median longitudinal furrow; POL = 5.5; OL = 3; OOL = 3; scutum shining or dull, weakly or strongly granulated; notauli complete, distinctly or almost separated posteriorly; scutellum and metanotum shining, smooth, without sculpture; propodeum dull, with a median longitudinal furrow on dorsal surface; posterior surface reticulate rugose; dorsal surface with two large smooth areas on sides of median furrow; fore wing hyaline, without dark transverse bands; stigmal vein with distal part much longer than proximal part; dorsal process of parameres (Fig. 48 A) very reduced, with distal apex rounded; maxillary palps 5-segmented; labial palps 2-segmented.

Distribution: rare in Sweden (only found in Sk.: Vitemölla, August, by M. Olmi). Not yet found in Denmark, Norway or East Fennoscandia. Spain, Algeria.

Biology: adults in pastures and fields in July. Hosts parasitized in August. A parasitoid of Delphacidae. Reared from *Kelisia sabulicola* W. Wagner.

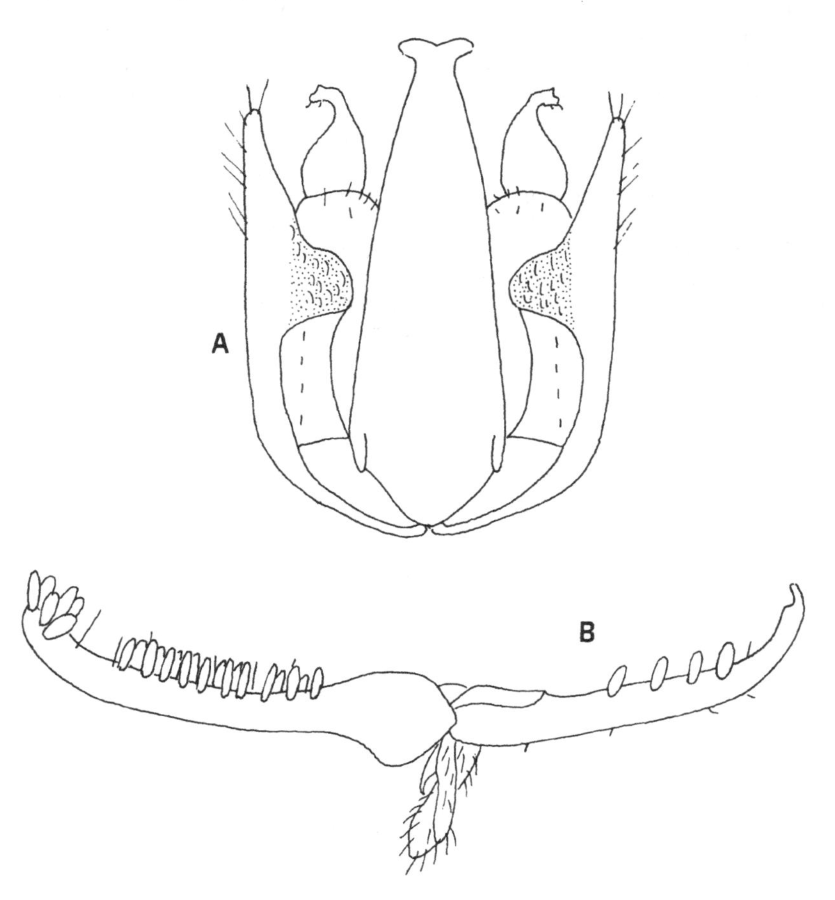

Fig. 48. *Gonatopus pallidus* (Ceballos): male genitalia (A) and chela (B).

25. ***Gonatopus formicicolus***
(Richards, 1939)
Plate 25, Figs 49-50.

Donisthorpina formicicola Richards, 1939: 201.
nec *Dicondylus pallidus* Ceballos, 1927: 107.

Female (Plate 25): apterous; length 3.00-3.18 mm; testaceous; antennae brown, with segments 1-3 yellow; petiole black; part of gaster brown; antennae thickened distally; antennal segments in following proportions: 7:5:11:7:5:5:4:4:4:7.5; head shining, excavated, without sculpture, except for front part of vertex and occiput granulated; pronotum shining, alutaceous, crossed by a strong transverse impression; pleura and posterior surface of propodeum transversely striate; meso-metapleural suture obsolete; metathorax + propodeum shining, without sculpture, except for striae on pleura and posterior surface; fore tarsal segments in following proportions: 12:2:4:12:18; enlarged claw (Fig. 49 B) with a subapical tooth and a row of 4-6 lamellae; segment 5 of fore tarsus (Fig. 49 B) with two rows of 17-22 lamellae; distal apex with a group of 6-12 lamellae; maxillary palps 5-segmented; labial palps 2-segmented.

Male (Fig. 50): fully winged; length 2.06-2.70 mm; black, with mandibles, tegulae, legs and antennal segments 1-2 yellow; antennae not thickened distally; antennal segment 3 more than four times as long as broad (4.33); antennal segments in following proportions: 6:5:13:10.5:10.5:9:9:9:9:11; head shining, haired, without sculpture or weakly punctate; frontal line absent; temples distinct; occipital carina invisible; POL = 6; OL = 2; OOL = 3; scutum haired, granulated; notauli complete, joined posteriorly; scutellum and metanotum shining, without sculpture or weakly punctate; propodeum reticulate rugose, without keels; fore wing hyaline, without dark transverse bands; dorsal process of parameres (Fig. 49 A) very short, distally pointed; maxillary palps 5-segmented; labial palps 2-segmented.

Distribution: rare in East Fennoscandia (only found in N: Tvärminne, 14.VII, by H. Lindberg); not yet found in Sweden, Norway or Denmark. Japan, Hungary, Netherlands, England, Wales, Italy.

Biology: adults in pastures and fields in July. A parasitoid of Delphacidae. In other European countries reared from *Javesella pellucida* (F.) and *dubia* (Kirschbaum); *Ribautodelphax pungens* (Ribaut); *Muellerianella fairmairei* (Perris); *Megadelphax sordidulus* (Stål); *Xanthodelphax stramineus* (Stål). The species may be parasitized by *Helegonatopus* sp. near *dimorphus* (Hoffer) (Hymenoptera Encyrtidae).

26. ***Gonatopus dromedarius***
(A. Costa, 1882)
Plate 26, Fig. 51.

Dicondylus dromedarius A. Costa, 1882: 38.
Gonatopus albosignatus Kieffer, 1904: 358.

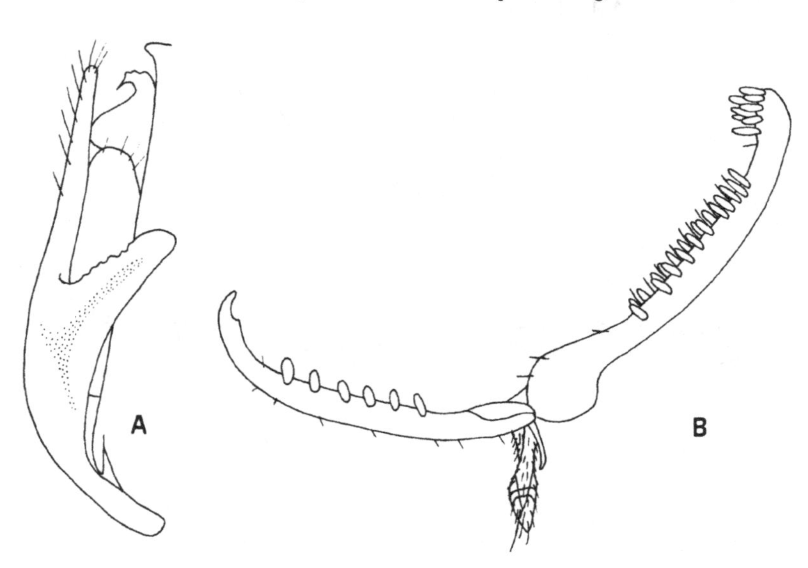

Fig. 49. *Gonatopus formicicolus* (Richards): male genitalia (right half removed) (A) and chela (B).

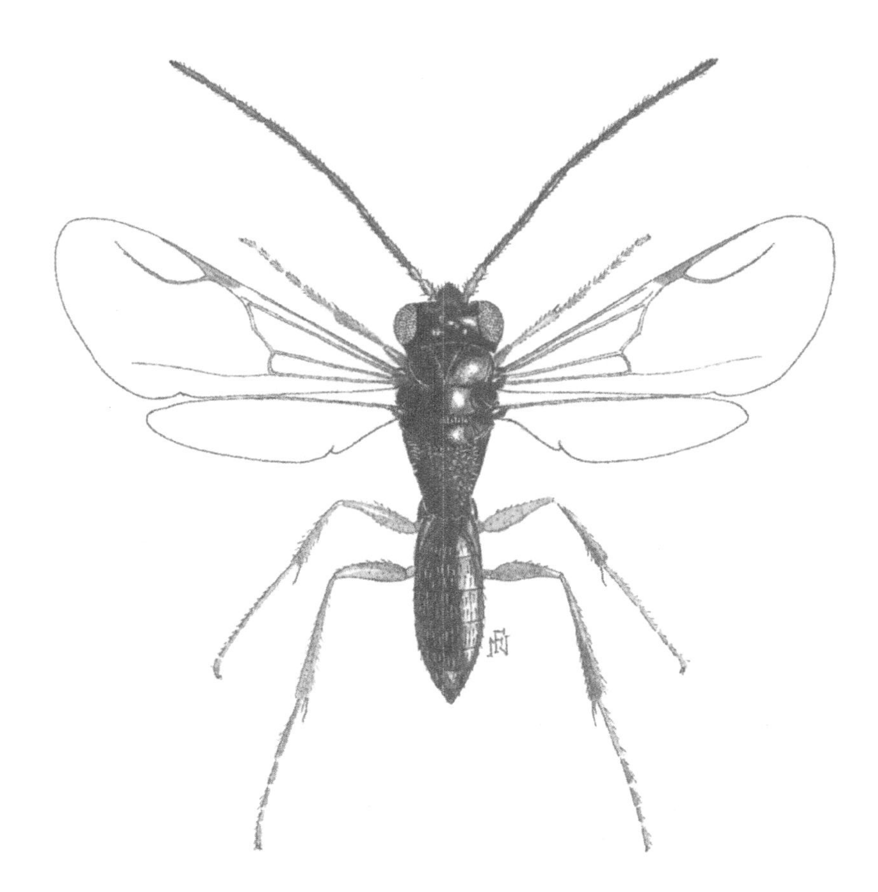

Fig. 50. Male of *Gonatopus formicicolus* (Richards).

Female (Plate 26): apterous; length 2.40-3.75 mm; colour very variable, from completely testaceous-yellow, with petiole black, to almost completely brown or black; specimens occur with intermediate colours (partly brown, or testaceous, or reddish); colour of antennae variable, from completely testaceous-yellow to almost completely black or brown; legs completely testaceous, or partly brown or black; antennae thickened distally; antennal segments in following proportions: 10:6:10:7:6:6:5.5:5.5:5.5:5:10; head excavated, shining or dull, without sculpture or slightly granulated; pronotum crossed by a strong transverse impression, shining or dull, without sculpture or weakly granulated; metanotum flat, hollow behind scutellum, transversely striate, short, at most as long as scutellum; sides of metanotum prominent, rounded or pointed; meso-metapleural suture distinct and complete; metathorax + propodeum shining, with posterior surface and pleura transversely striate; anterior surface smooth and unsculptured; disc of metathorax + propodeum with or without trace of a median longitudinal furrow; segment 1 of fore tarsus longer or shorter than segment 4; enlarged claw (Fig. 51 B) with a large subapical tooth and a row of 7-11 lamellae; segment 5 of fore tarsus (Fig. 51 B) with two rows of 11-23 lamellae; distal apex with a group of approximately 7-23 lamellae; maxillary palps with 3-4 segments; labial palps 2-segmented.

Male: fully winged; length 1.68-2.00 mm; black; clypeus and mandibles whitish or testaceous; antennae brown; legs brown, with joints and tarsi testaceous; antennae not thickened distally, with segment 3 less than three and a half times as long as broad (6:2); antennal segments in following proportions: 5:4:6:5.5:5.5:5:5:5.5:5:7; head dull, granulated, with two shining and smooth oval areas situated between posterior ocelli and eyes; these ar-

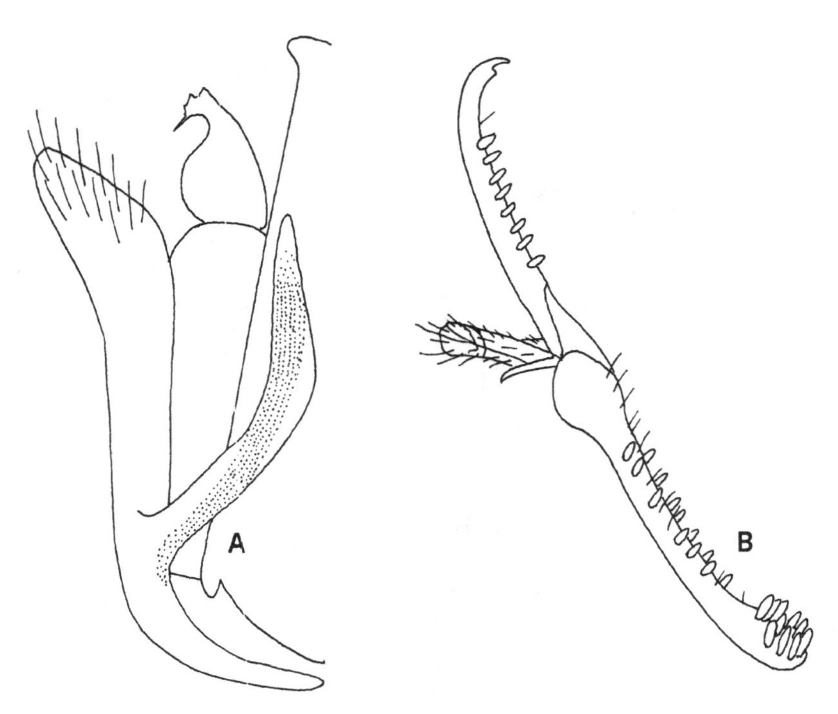

Fig. 51. *Gonatopus dromedarius* (A. Costa): male genitalia (right half removed) (A) and chela (B).

eas surrounded anteriorly by a strong and high carina (see fig. 56 C); occipital carina absent; POL = 5; OL = 2.5; OOL = 3; scutum dull, granulated; notauli incomplete, reaching approximately 0.6 length of scutum; occasionally notauli weakly visible even in posterior half of scutum and then complete and joined; scutellum dull, granulated; metanotum shining, smooth, without sculpture; propodeum dull, rugose; fore wing hyaline, without dark transverse bands, with veins weakly visible (apparently only costal cell completely enclosed by pigmented veins); dorsal process of parameres (Fig. 51 A) long and slender; maxillary palps with 3-4 segments; labial palps 2-segmented.

Distribution: rare in Sweden (only found in Sk.: Vitemölla, 26.VI.1948, by Bo Tjeder). Not yet found in Denmark, Norway or East Fennoscandia. Widespread in South Europe (France. Italy, Spain, Greece); apparently scarce in Central Europe (Netherlands, Hungary). Also in Turkey, Israel, Egypt, Algeria, Canary Islands.

Biology: adults in pastures and fields. A parasitoid of Delphacidae (in South Europe also of Issidae and Flatidae). Reared from *Laodelphax striatellus* (Fallén); *Dicranotropis hamata* (Boheman); *Megadelphax* sp.

27. ***Gonatopus distinctus*** Kieffer, 1906

Plate 27, Fig. 52.

Gonatopus distinctus Kieffer in Kieffer & Marshall, 1906: 509.
Gonatopus septemdentatus J. Sahlberg, 1910: 11.

Female (Plate 27): apterous; length 3.00-4.00 mm; head black, with mandibles, clypeus and anterior part of frons yellow; occasionally also occiput yellow; antennae black or brown, with segments 1-2 yellow; occasionally also antennal segment 10 yellow; thorax, propodeum, petiole and gaster black; occasionally scutum partly yellow; legs yellow, with clubs of fore femora brownish; occasionally also clubs of mid and hind femora and mid and hind coxae brownish; head excavated, shining, granulated; pronotum crossed by a strong transverse impression, unhaired, weakly alutaceous; scutum dull, rugose; metanotum transversely striate, laterally protruding; lateral protrusions rounded; meso-metapleural suture distinct and complete; mesopleura, metapleura and posterior surface of propodeum transversely striate; ante-

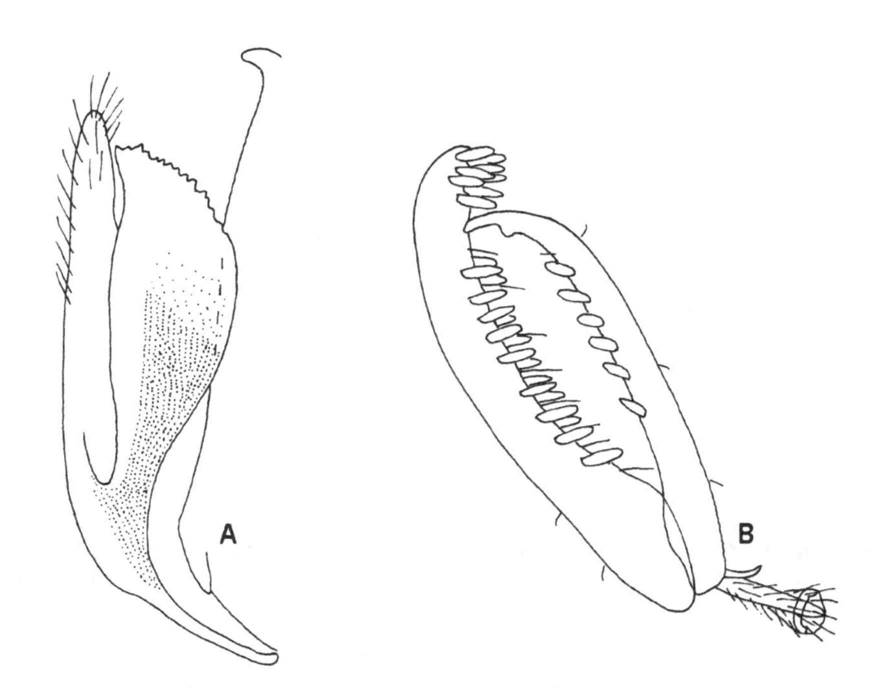

Fig. 52. *Gonatopus distinctus* Kieffer: male genitalia (right half removed) (A) and chela (B).

rior surface of metathorax + propodeum weakly rugose, with trace of a median longitudinal furrow; fore tarsal segments in following proportions: 15:2:4:11:18; enlarged claw (Fig. 52 B) with a large subapical tooth and a row of 5-13 lamellae; segment 5 of fore tarsus (Fig. 52 B) with 2 rows of 16-27 lamellae; distal apex with a group of 5-25 lamellae; maxillary palps 4-segmented; labial palps 2-segmented.

Male: fully winged; length 2.50-2.62 mm; head black, with mandibles testaceous; antennae brown; thorax and propodeum black; gaster brown-reddish; legs brown, with light tarsi; antennae not thickened distally, with segment 3 four or more than four times as long as broad; head shining, granulated; frontal line absent; temples distinct; occipital carina absent; scutum shining, granulated; notauli complete, usually separated posteriorly, rarely joined; minimum distance between notauli shorter than breadth of ocelli; scutellum and metanotum shining, smooth, weakly punctate; propodeum reticulate rugose, shining, without longitudinal keels on posterior surface; dorsal process of parameres (Fig. 52 A) long, with apical and inner margin serrate; maxillary palps with 3-4 segments; labial palps 2-segmented.

Distribution: rare in Denmark (found only at SJ: Sandager) and Norway (only found in Bø (Tofteholmen, 28.V-7.VII.1991, by Lars Ove Hansen) and Ry (Sevheimsheia, Finnoy, 1-16.7.1992, by John Skartveit)); rather common in Sweden (Bl., G. Sand., Vstm.) and East Fennoscandia (Al, N, Ta, Sa, Oa, Vib, Kr). Widespread in Europe, but always uncommon; rather common in England. Also in Mongolia.

Biology: adults in pastures and fields from May to July. A parasitoid of Delphacidae. Reared from *Dicranotropis hamata* (Boheman); *Gravesteiniella boldi* (Scott); *Ditropis pteridis* (Spinola); *Hyledelphax elegantulus* (Boheman); *Javesella pellucida* (F.) and *discolor* (Boheman); *Kosswigianella exigua* (Boheman); *Ribautodelphax angulosus* (Ribaut). Mating observed by Waloff (1974). The number of generations in Fennoscandia and Denmark is not known, but in England the species is univoltine (Waloff, 1974, 1975). It overwinters in a cocoon spun on the host plant. The species may be parasitized by *Helegonatopus rasnitzyni* Tryapitsyn (Hymenoptera Encyrtidae) (Tryapitsyn, 1978). Information on biology in Waloff (1974, 1975).

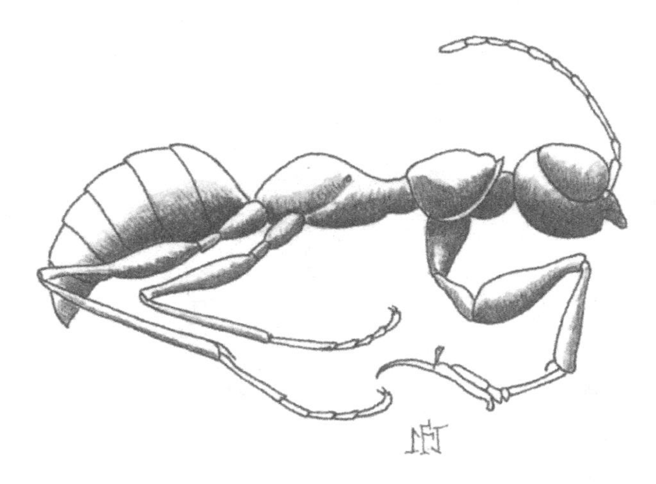

Fig. 53. Female of *Gonatopus pedestris* Dalman in lateral view.

28. ***Gonatopus pedestris*** Dalman, 1818

Plate 28, Figs 53-54.

Gonatopus pedestris Dalman, 1818: 86.
Gonatopus ljunghii Westwood, 1833: 496.
Gonatopus leucostomus J. Sahlberg, 1910: 14.

Female (Plate 28, Fig. 53): apterous; length 2.82-3.00 mm; head black or brown, with mandibles, clypeus and anterior part of frons yellow or yellow-brown; antennae yellow or yellow-brown; thorax, propodeum and gaster black; occasionally pronotum testaceous-brown; legs yellow, occasionally with part of coxae and part of femoral clubs brown or black; head weakly excavated, shining, granulated; temples distinct; occipital carina absent; pronotum shining, weakly granulated, not crossed by a strong transverse impression or very weakly impressed; scutum dull, granulated; metanotum with a semicircular keel; metathorax + propodeum granulated, usually with trace of a median longitudinal furrow; posterior surface of propodeum completely transversely striate; meso-metapleural suture obsolete; enlarged claw (Fig. 54 B) with a small subapical tooth and with a row of a few bristles; segment 5 of fore tarsus (Fig. 54 B) with 1 row of 2-7 minute lamellae situated in distal half; distal apex with a group of 4-9 lamellae; maxillary palps with 2-4 segments; labial palps 2-segmented.

Male: fully winged; length 2.40-2.93 mm; head black, with mandibles yellow; antennae black-brown, with segments 1-2 partly yellow; thorax and propodeum black; gaster brown-black; legs brown-black, with stalk of fore femora and fore tibiae yellow; antennae not thickened distally, with segment 3 more than four times as long as broad (5.6); temples distinct; occiput concave; occipital carina absent; POL = 7; OL = 3; OOL = 3; OOL longer than breadth of ocelli (3:2); scutum dull, granulated; notauli complete, separated posteriorly; scutellum and metanotum granulated; propodeum reticulate rugose; fore wing hyaline, without dark transverse bands; dorsal process of parameres (Fig. 54 A) long, slender, with distal apex pointed; maxillary palps with 2-4 segments; labial palps 2-segmented.

Distribution: rather common in Denmark (F, NEZ), Sweden (Sk., Vg., Hls.) and East Fennoscandia (Al, Ab, N, St, Sa, ObS, Vib, Kr). Probably rather common in Norway, but so far found only in Ø (Kirkoen, Hvaler, V.1926, by Munster). Widespread, but always uncommon, in Central and Eastern Europe; also in Siberia, Turkey.

Biology: adults in pastures and fields from May to August. A parasitoid of Cicadellidae Macrostelini. Reared from *Macrosteles laevis* (Ribaut) and *quadripunctulatus* (Kirschbaum).

29. ***Gonatopus lunatus*** Klug, 1810

Plate 29, Figs 55-56.

Gonatopus lunatus Klug, 1810: 164.
Gonatopus gracilicornis Kieffer, 1904: 361.

Female (Plate 29, Fig. 55): apterous; length 2.31-3.56 mm; head reddish-yellow, with or without a brown band on vertex; antennae brown, with segments 1-2 yellow; pronotum testaceous-reddish, occasionally with transverse impression and most

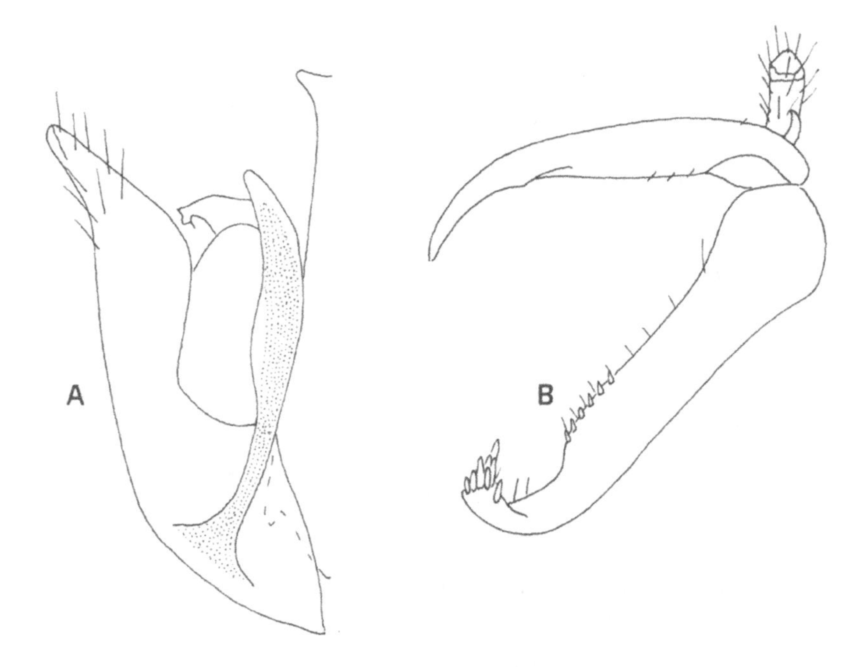

Fig. 54. *Gonatopus pedestris* Dalman: male genitalia (right half removed) (A) and chela (B).

of disc brown; scutum yellow, occasionally with anterior and posterior margins brown; scutellum and metathorax + propodeum black, occasionally reddish-yellow; gaster brown or black, with reddish nuances; legs reddish-yellow, occasionally with brownish coxae and femoral clubs; body rarely completely testaceous-yellow, with black petiole; head excavated, shining, without sculpture, except for occiput and anterior part of frons granulated; antennae slender, more than 3,5 times as long as head; pronotum crossed by a strong transverse impression, shining, granulated; scutum shining, with

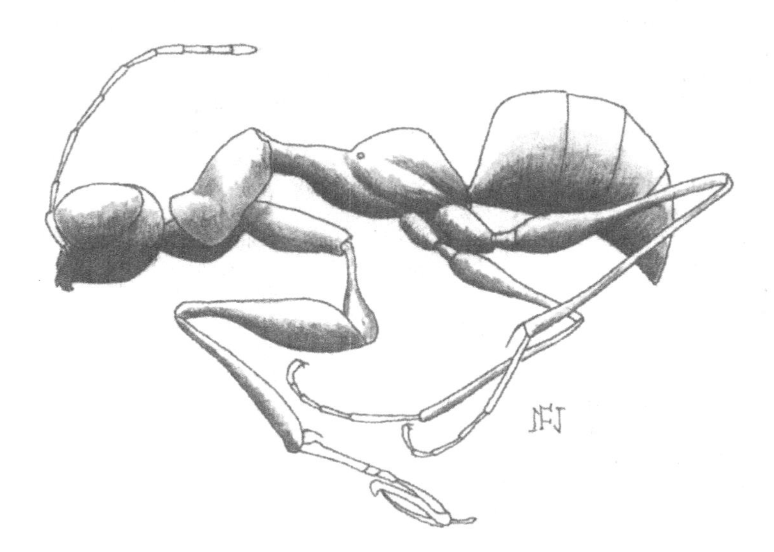

Fig. 55. Female of *Gonatopus lunatus* Klug in lateral view.

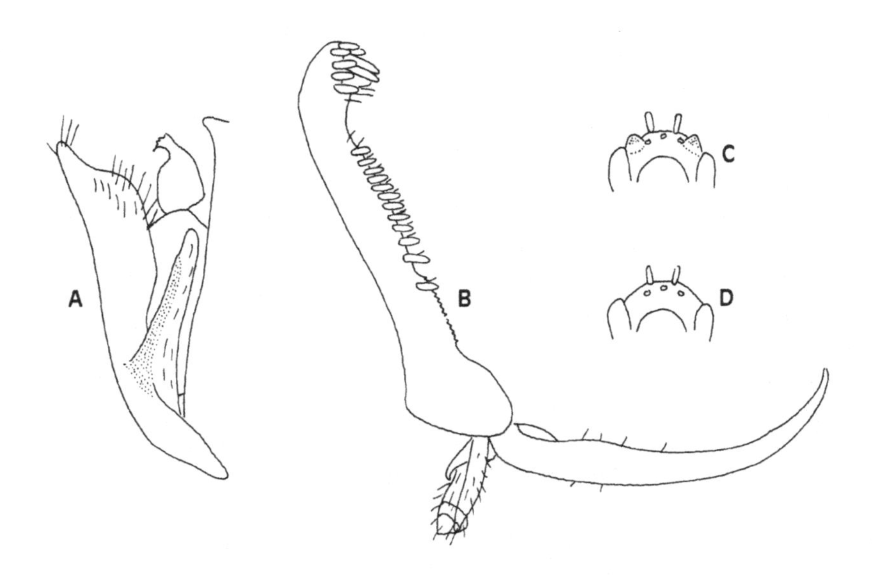

Fig. 56. *Gonatopus lunatus* Klug: male genitalia (right half removed) (A), chela (B) and male head (in posterior view) (C); D: head of male of *Gonatopus clavipes* (Thunberg) in posterior view.

a few longitudinal keels; metanotum flat, weakly transversely striate; metathorax + propodeum dull or shining, strongly or weakly granulated, with trace of a median furrow; this trace occasionally very deep; posterior surface of propodeum transversely striate; meso-metapleural suture obsolete; fore tarsal segments in following proportions: 13:2:4:13:21; enlarged claw (Fig. 56 B) with a small subapical tooth and a row of 4-7 peg-like hairs or bristles; segment 5 of fore tarsus (Fig. 56 B) with inner side serrate proximally and with 2 rows of 1-3 + 12-17 lamellae (shortest row proximal, not extending beyond 0.5 length of segment); distal apex with a group of 9-14 lamellae; maxillary palps 4-segmented; labial palps 2-segmented.

Male: fully winged; length 1.75-1.87 mm; black, with mandibles testaceous; tegulae yellow; legs banded; antennae not thickened distally, with segment 3 less than 3.5 times as long as broad; antennal segments in following proportions: 4.5 : 4 : 6 : 5.5 : 5.5 : 5.5 : 5.5 : 5.5 : 5.5 : 7.5; head dull, haired, granulated, with a shining, smooth oval area between posterior ocelli and eyes; each area surrounded anteriorly by a very prominent apophysis (Fig. 56 C); occipital carina absent; temples distinct; scutum dull, granulated; notauli complete, separated posteriorly; minimum distance between notauli much shorter than antennal segment 2; scutellum dull, granulated; metanotum shining, without sculpture; propodeum weakly reticulate rugose, without keels; dorsal surface of propodeum with a weak median furrow; fore wing hyaline, without dark transverse bands; dorsal process of parameres (Fig. 56 A) slender, with distal apex pointed; maxillary palps 4-segmented; labial palps 2-segmented.

Distribution: common in Denmark (NEJ, F, NEZ), Sweden (Sk., Öl., Nrk., Sdm., Upl.) and East Fennoscandia (Ab, N, St., Ta, Sa, Vib). Probably common in Norway, but so far found only in Bø (Underlia, 1-30.VI.92, by Lars Ove Hansen) and VE (Langøya, 28.V-8.VII.91, by Lars Ove Hansen). Widespread in Europe, also in Korea, Mongolia, Siberia, Kazakhstan, Turkmenistan, Turkey, Lebanon, Israel, Morocco, Azores Islands, Madeira.

Biology: adults in pastures and fields from June to August. A parasitoid of Cicadellidae Deltocephalinae. Reared from *Psammotettix alienus* (Dahlbom); *Euscelis incisus* (Kirschbaum); *Errastunus* sp.; *Arthaldeus pascuellus* (Fallén); *Macrosteles* sp.; *Adarrus multinotatus* (Boheman). Information on behaviour and biology by Lindberg (1950).

30. ***Gonatopus striatus*** Kieffer, 1905

Plate 30, Fig. 57.

Gonatopus striatus Kieffer in Kieffer & Marshall, 1905: 92.

Female (Plate 30): apterous; length 3.00-4.00 mm; head brown or black, with mandibles, clypeus and anterior part of frons testaceous; occiput testaceous-reddish; antennae black, with segments 1-3 yellow; thorax, propodeum, petiole and gaster black; legs testaceous, occasionally with clubs of fore femora brownish; head excavated, shining, granulated; pronotum crossed by a strong transverse impression, shining, without sculpture, finely haired; scutum shining, with a few longitudinal keels, laterally with two pointed prominences; metathorax + propodeum shining, without sculpture, except for pleura and posterior surface of propodeum transversely striate; meso-metapleural suture complete or only proximally distinct; segment 1 of fore tarsus slightly shorter than segment 4 (10:12); enlarged claw (Fig. 57 B, C) with a small subapical tooth and a row of 5-7 peg-like hairs or lamellae; segment 5 of fore tarsus (Fig. 57 C) with inner side proximally serrate, with 2-3 rows of 24-40 lamellae extending beyond 0.5 length of segment; distal apex with a group of 6-40 lamellae; maxillary palps with 4-5 segments; labial palps with 2-3 segments; palpal formula: 4/3, 5/2, 5/3.

Male: fully winged; length 1.69-2.37 mm; black; mandibles black, with distal half testaceous; legs brown, with fore tibiae testaceous; antennae not thickened distally, with segment 3 less than 3.5 times as long as broad; antennal segments in following proportions: 6:4.5:6.5:7:7:7:7:7:7:10; head dull, granulated and reticulate rugose; vertex with a shining, smooth oval area between posterior ocelli and eyes; occipital carina absent; POL = 6; OL = 3; OOL = 5; temples distinct; occiput concave; scutum dull, completely granulated; notauli complete, separated posteriorly; minimum distance between notauli shorter than antennal segment 2; scutellum dull, completely granulated; metanotum shining, smooth, without sculpture; propodeum strongly reticulate rugose, without longitudinal keels; fore wing hyaline, without dark transverse bands; dorsal process of parameres (Fig. 57 A) very long, slender; maxillary palps 5-segmented; labial palps with 2-3 segments.

Distribution: rather common in Denmark (F, NWZ, NEZ), Sweden (Sk., Bl., Sm., Öl., Gtl., Ög., Nrk., Upl.) and East Fennoscandia (Ab, N). Probably rather common in Norway, but so far found only in Bø (Kinnartangen, 2-28.V.1991, by Lars Ove Hansen) and AK (Bukten, Bygdø, V.1934, by

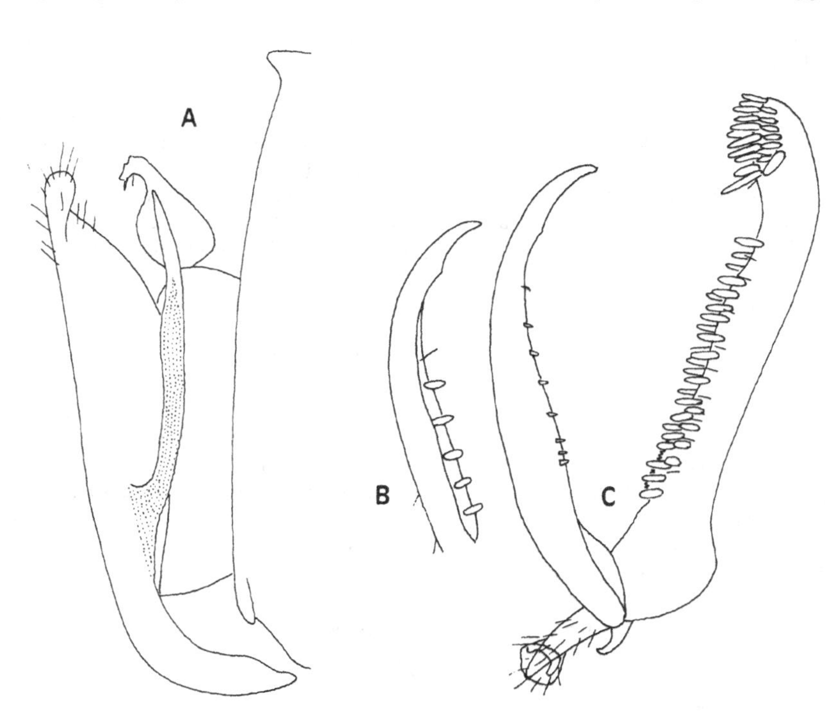

Fig. 57. *Gonatopus striatus* Kieffer: male genitalia (right half removed) (A), enlarged claw with lamellae (B) and chela (C: enlarged claw with peg-like hairs).

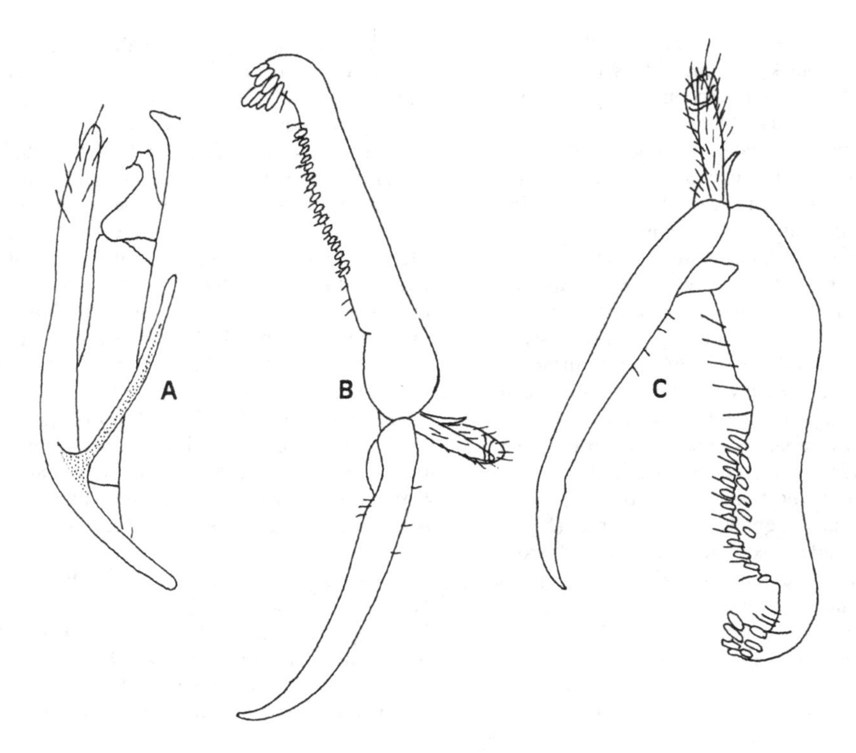

Fig. 58. *Gonatopus formicarius* Ljungh: male genitalia (right half removed) (A) and chela (B); C: chela of *Gonatopus spectrum* (Vollenhoven)

Munster). Widespread in Europe, but always uncommon. Also in Mongolia, Turkey.

Biology: adults in pastures and fields from May to July. A parasitoid of Cicadellidae Aphrodinae. Reared from *Aphrodes bicincta* (Schrank).

31. ***Gonatopus spectrum***
(Van Vollenhoven, 1874)
Plate 31, Fig. 58.

Dryinus spectrum Van Vollenhoven, 1874: 159.

Female (Plate 31): apterous; length 2.80-3.12 mm; testaceous- yellow, with petiole black and gaster brownish; rarely metathorax + propodeum dark; occasionally with two brown spots on sides of propodeum; head excavated, without sculpture, shining, unhaired; pronotum crossed by a strong transverse impression, shining, smooth; metanotum shining, transversely striate; metathorax + propodeum shining, without sculpture, except for pleura and posterior surface of propodeum transversely striate; meso- metapleural suture obsolete; segment 1 of fore tarsus as long as segment 4; enlarged claw (Fig. 58 C) with a small subapical tooth and a row of 4-5 peg-like hairs or bristles; segment 5 of fore tarsus (Fig. 58 C) with inner side usually not serrate proximally (or occasionally serrate), with 2 rows of 15-17 lamellae situated on a distinct prominence in distal half of segment; distal apex with a group of 7-10 lamellae; maxillary palps with 4-5 segments; labial palps 2-segmented.

Male: unknown.

Distribution: rare in Sweden (only found in Sk. (Kullaberg, 25.VI) and in Hall. (Tönnersa, 9.IX)). Not yet found in Denmark, Norway or East Fennoscandia. Widespread in Europe, but always rare: Hungary, Bulgaria, Germany, Austria, Netherlands, Switzerland, France, North Italy, Greece.

Biology: adults in pastures and fields from June to September. A parasitoid of Cicadellidae Deltocephalinae. Reared from *Jassargus* sp.; *Adarrus* sp.; *Rhopalopyx* sp.

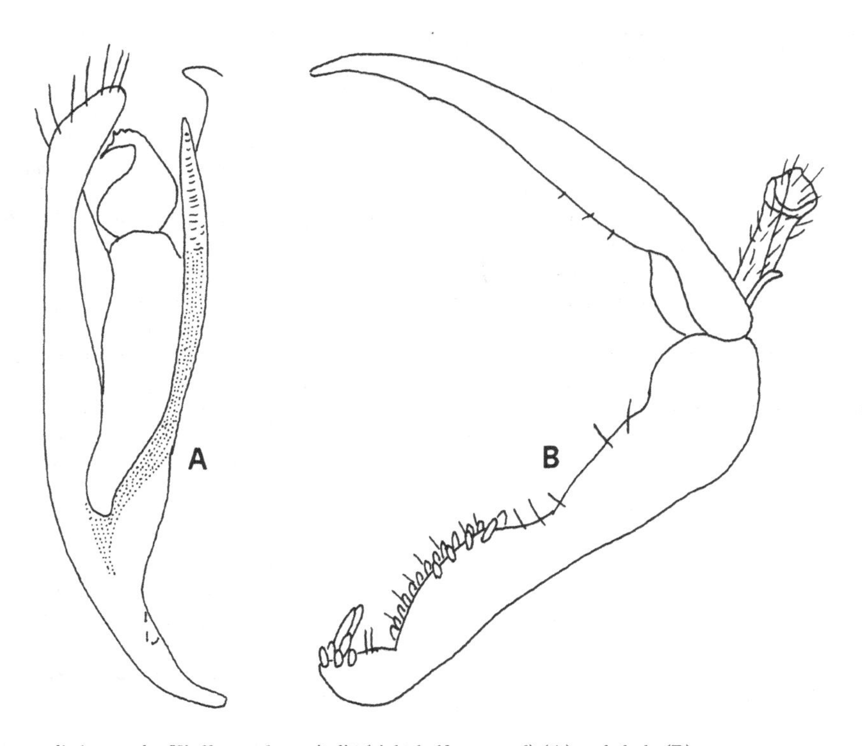

Fig. 59. *Gonatopus distinguendus* Kieffer: male genitalia (right half removed) (A) and chela (B).

32. ***Gonatopus distinguendus*** Kieffer, 1905

Plates 32-33, Fig. 59.

Gonatopus flavicornis Thomson, 1860: 181 (preoccupied).
Gonatopus distinguendus Kieffer in Kieffer & Marshall, 1905: 116.
Gonatopus excavatus J. Sahlberg, 1910: 9.
Gonatopus thomsoni Hellén, 1953: 94.

Female (Plates 32-33): apterous; length 2.68-3.75 mm; colour very variable from black to reddish-black, with intermediate colours; head black, or reddish, or testaceous-reddish, with brown or black bands or spots; antennae reddish-yellow, with segment 10 brownish; pronotum completely black or reddish, with brown spots; scutum black, or yellow, or reddish, or brown; scutellum and metathorax + propodeum black; gaster black or reddish, with first and last tergites black; legs testaceous, with partly brown femoral clubs and coxae; head excavated, shining, unhaired, weakly granulated; pronotum crossed by a strong transverse impression, dull, granulated; scutum shining, without sculpture or granulated; metanotum transversely striate; meso-metapleural suture obsolete; metathorax + propodeum dull, granulated; mesopleura and metapleura not or weakly transversely striate; posterior surface of propodeum weakly transversely striate; segment 1 of fore tarsus as long as segment 4; enlarged claw (Fig. 59 B) with a small subapical tooth and a row of 3-6 peg-like hairs; segment 5 of fore tarsus (Fig. 59 B) with inner side serrate proximally (occasionally not serrate) and with two rows of 12-21 lamellae situated on a distinct prominence in distal half of segment; distal apex with a group of 3-11 lamellae; maxillary palps with 4-5 segments; labial palps 2-segmented.

Male: fully winged; length 1.93-2.62 mm; black; mandibles testaceous; antennae and legs banded; antennae not thickened distally, with segment 3 less than three times as long as broad; antennal segments in following proportions: 3.5:4.5:7.5:7:6.5:7:6.5:6.5:6.5:9; head dull, granulated; frontal line absent; occipital carina absent; POL = 5; OL = 2.5; OOL = 4; scutum, scutellum and metanotum dull, granulated; notauli complete, separated posteriorly; minimum distance between notauli approximately as long as antennal segment 2; propodeum reticulate rugose, with a median longitudinal furrow on dorsal surface; fore wing hyaline, without dark transverse bands; dorsal process of parameres (Fig. 59 A) slender, with distal apex pointed; maxillary palps 4-segmented; labial palps 2-segmented.

Distribution: rather common in Denmark (F, NEZ), Sweden (Sk., Hall., Öl., Gtl., Ög.) and East Fennoscandia (Al, Ab, N, Vib). Not yet found in Norway. Widespread in Europe, but always rather common. Also in Mongolia, Siberia, Turkey, Iran.

Biology: adults in pastures and fields from May to September. A parasitoid of Cicadellidae Deltocephalinae. Reared from *Psammotettix cephalotes* (H.-S.) and *confinis* (Dahlbom); *Mocuellus collinus* (Boheman).

33. ***Gonatopus formicarius*** Ljungh, 1810

Plate 34, Fig. 58.

Gonatopus formicarius Ljungh, 1810: 162.
Gonatopus formicinus Krogerus, 1932: 124 (nomen nudum).

Female (Plate 34): apterous; length 2.24-3.40 mm; head black, with occiput, clypeus, mandibles and anterior part of frons testaceous; antennae black, with segments 1-2 yellow; pronotum black or brown, with disc and anterior margin reddish; scutum, scutellum and metathorax + propodeum black, with distal apex of propodeum testaceous; occasionally scutum partly yellow; petiole and gaster black, with tergite 1 yellow; legs yellow, with coxae and clubs of femora brownish; occasionally legs completely yellow; antennae thickened distally; antennal segments in following proportions: 6:5:10:5:4.5:5:4.5:4.5:4:6; head more or less excavated, dull, granulated; POL = 2; OL = 2; OOL = 7; pronotum crossed by a strong transverse impression, shining, granulated; scutum dull, with a few longitudinal keels; metanotum not striate transversely; metathorax + propodeum dull, granulated, rarely with a strong median longitudinal furrow on disc; mesopleura, metapleura and posterior surface of propodeum not striate transversely; meso-metapleural suture distinct and complete; segment 1 of fore tarsus slightly longer than segment 4; enlarged claw (Fig. 58 B) with a small subapical tooth and a row of a few hairs or bristles; segment 5 of fore tarsus (Fig. 58 B) with inner side not serrate proximally and with 1 row of 10-20 lamellae; rarely with 2 rows of lamellae; distal apex with a group of 7-14 lamellae; maxillary palps 5-segmented; labial palps 2-segmented.

Male: fully winged; length 1.56-1.87 mm; black; mandibles testaceous; antennae brown; legs brown, with fore tibiae and fore tarsi yellow; antennae not thickened distally, with segment 3 less than three and a half times as long as broad (6:2); antennal segments in following proportions: 5:4:6:6:6:6:5.5:6:5.5:8; head shining, granulated; frontal line absent; occipital carina absent; POL = 5, OL = 2,5; OOL = 4; scutum dull, granulated; notauli complete, separated posteriorly; minimum distance between notauli shorter than antennal segment 2; scutellum dull, granulated; metanotum shining, without sculpture; propodeum dull, reticulate rugose, with a median longitudinal furrow on dorsal surface; fore wing hyaline, without dark transverse bands; dorsal process of parameres (Fig. 58 A) slender and with distal apex pointed; maxillary palps 5-segmented; labial palps 2-segmented.

Distribution: uncommon in Denmark (NWJ, NEJ, NEZ, B) and East Fennoscandia (Ab, Ta, Vib); rather common in Sweden (Sk., Bl., Hall., Sm., Öl., Gtl., Nrk., Upl., T. Lpm.); not yet found in Norway. Widespread in Europe, but always uncommon; also in Siberia, Uzbekistan, Turkey.

Biology: adults in pastures and fields from May to September. A parasitoid of Cicadellidae Deltocephalinae. Reared from *Verdanus abdominalis* (F.); *Psammotettix alienus* (Dahlbom), *confinis* (Dahlbom) and *cephalotes* (H.-S.); *Adarrus multinotatus* (Boheman).

34. ***Gonatopus clavipes*** (Thunberg, 1827)

Plates 35-36, Fig. 60.

Gelis clavipes Thunberg, 1827: 202.
Gonatopus sepsoides Westwood, 1833b: 496.
Gonatopus pilosus Thomson, 1860: 180.
Gonatopus borealis J. Sahlberg, 1910: 12.
Gonatopus krogerii Hellén, 1935: 8 (nomen nudum).
Gonatopus barbatellus Richards, 1939: 213.

Female (Plate 35): apterous; length 2.18-3.75 mm; colour very variable; usually head black, with mandibles, clypeus and anterior region of frons yellow; occasionally occiput yellow; antennae black, with segments 1-2 and part of 3 yellow; mesosoma black; occasionally pronotum almost completely reddish or with sides reddish and transverse impression black; occasionally scutum yellow or dark-reddish; occasionally metathorax + propodeum dark-reddish; gaster black, with reddish nu-

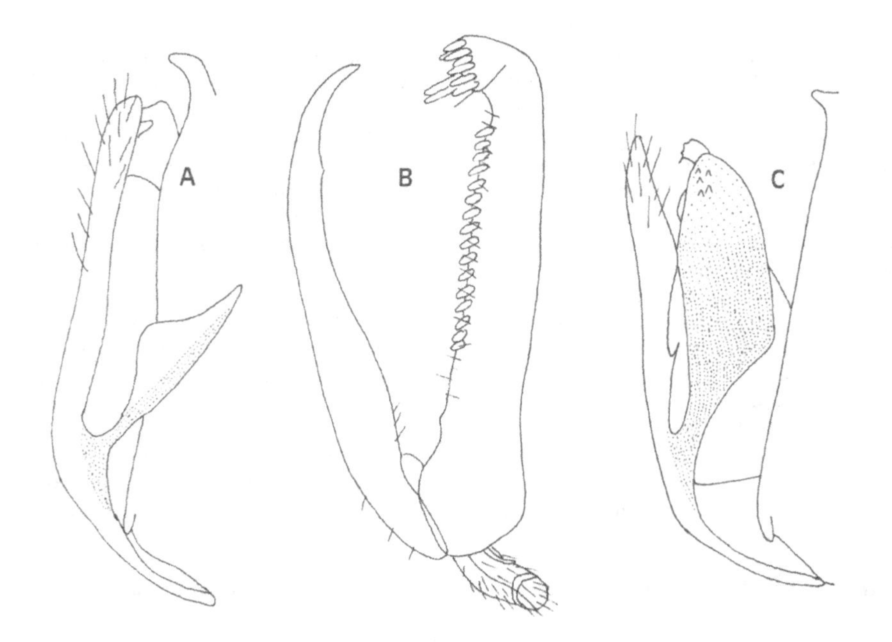

Fig. 60. *Gonatopus clavipes* (Thunberg): male genitalia (right half removed) (A) and chela (B); C: male genitalia (right half removed) of *Gonatopus albifrons* Olmi.

ances; occasionally gaster almost completely reddish; legs reddish-yellow, with femoral clubs brownish; head excavated, shining or dull, granulated or alutaceous; pronotum crossed by a strong transverse impression, granulated, shining or dull, haired or unhaired; scutum granulated, without lateral points, with a few longitudinal keels; metanotum transversely striate; meso-metapleural suture distinct and complete; metathorax + propodeum shining or dull, haired, with pleura and posterior surface of propodeum transversely striate, granulated or without sculpture among striae; disc of metathorax + propodeum with or without a median longitudinal furrow; fore tarsal segments in following proportions: 12:2:4:14:21; enlarged claw (Fig. 60 B) with a small subapical tooth and a row of a few hairs, or bristles, or peg-like hairs; segment 5 of fore tarsus (Fig. 60 B) with inner side not serrate proximally, with 1-2 rows of 11-23 lamellae; distal apex with a group of 7-17 lamellae; maxillary palps 5-segmented; labial palps 2-segmented.

Male (Plate 36): fully winged; length 1.68-2.75 mm; black, with mandibles yellow; antennae brown-black; tegulae yellow; legs yellow-brown, with coxae dark; antennae not thickened distally, with segment 3 less than 3,5 times as long as broad; antennal segments in following proportions: 4:4:6:5.5:5.5:5:5:5:5.5:8; head dull, haired, granulated, with short irregular keels on frons and vertex, with a shining, smooth, oval area between posterior ocelli and eyes (Fig. 56 D); temples distinct; occipital carina absent; scutum dull, granulated; notauli complete, separated posteriorly; rarely notauli incomplete, reaching approximately 0.65 length of scutum; scutellum dull, granulated; metanotum shining, granulated or without sculpture; propodeum dull, reticulate rugose, with a median longitudinal furrow on dorsal surface; dorsal process of parameres (Fig. 60 A) transverse, short; maxillary palps 5-segmented; labial palps 2-segmented.

Distribution: common in Denmark (F, LFM, SZ, NEZ), Norway (but found only in AK, Bo, Bv, VE, TEi, Ry, Fn), Sweden (Sk., Bl., Hall., Sm., Gtl., Ög., Nrk., Sdm., Upl., Vstm., Dlr., Hrj., Vb., Ås. Lpm.) and East Fennoscandia (Al, Ab, N, St, Ta, Sa, Oa, Tb, Kb, Ok, ObS, Ks, Vib, Lr). This is the most common gonatopodine species in Fennoscandia and Denmark. Widespread in Europe, also in Korea, Mongolia, Siberia, Kazakhstan, Uzbekistan, Turkey, Iran, Azores Islands, Canary Islands.

Biology: adults in pastures and fields from May to

September. A parasitoid of Cicadellidae Deltocephalinae. Reared from *Psammotettix confinis* (Dahlbom), *alienus* (Dahlbom), *putoni* (Then), *nodosus* (Ribaut) and *cephalotes* (H.-S.); *Turrutus socialis* (Flor); *Sorhoanus xanthoneurus* (Fieber) and *assimilis* (Fallén); *Arocephalus languidus* (Flor), *punctum* (Flor) and *longiceps* (Kirschbaum); *Euscelis incisus* (Kirschbaum); *Arthaldeus striifrons* (Kirschbaum) and *pascuellus* (Fallén); *Verdanus abdominalis* (F.); *Jassargus flori* (Fieber) and *distinguendus* (Flor); *Errastunus ocellaris* (Fallén); *Cicadula* sp.; *Elymana sulphurella* (Zetterstedt); *Streptanus sordidus* (Zetterstedt); *Thamnotettix confinis* (Zetterstedt). Information on biology by Abdul-Nour (1976), Teodorescu (1982) and Waloff (1974, 1975). The number of generations in Fennoscandia and Denmark is not known, but the species is either bi- or trivoltine in Britain. It overwinters in a cocoon spun on the host plant (Waloff, 1974, 1975).

35. *Gonatopus albifrons* Olmi, **nom. n.**

Fig. 60.

nec *Gonatopus albosignatus* Kieffer, 1904: 358.
Labeo albosignatus Kieffer in Kieffer & Marshall, 1905: 224 (preoccupied; *Labeo* Haliday is junior synonym of *Gonatopus* Ljungh).

Female: unknown

Male: fully winged; length 1.68-1.84 mm; black or brown, with ventral side of head, clypeus, mandibles, anterior surface and lateral regions of frons (especially along orbits) testaceous-yellow; antennae not thickened distally, with segment 3 less than three times as long as broad; antennal segments in following proportions: 5:5:7:7:6.5:6.5:6:6:6:9; head dull, granulated; frontal line absent; occipital carina absent; POL = 7; OL = 2; OOL = 3; vertex with a shining, smooth oval area between posterior ocelli and eyes, which is surrounded anteriorly by a prominent apophysis (see fig. 56 C); scutum, scutellum and metanotum granulated; notauli complete, separated posteriorly; minimum distance between notauli shorter than breadth of ocelli (1:2); propodeum reticulate rugose, except for a small smooth surface near metanotum; dorsal surface with a deep median furrow; fore wing hyaline, without dark transverse bands; dorsal process of parameres (Fig. 60 C) long, broadened; maxillary palps 4-segmented; labial palps 2-segmented.

Distribution: rare in Sweden (found only at two localities in Sk.: Åhus, 10.VII; Vitemölla, 14.VI). Not yet found in Denmark, Norway or East Fennoscandia. Rare also in Hungary, Switzerland, France, Spain, Italy.

Biology: adults in June and July in pastures and fields. A parasitoid of Delphacidae. Reared in France from *Ribautodelphax imitans* (Ribaut).

Nomina nuda

borealis (Prenanteon) Hellén, 1935: 7.
flaviventre (Anteon) Hellén, 1935: 7.
formicinus (Gonatopus) Krogerus, 1932: 124 (see *Gonatopus formicarius* Ljungh).
krogerii (Gonatopus) Hellén, 1935: 8 (see *Gonatopus clavipes* (Thunberg)).
longicornis (Anteon) Hellén, 1935: 7.
nitidulus (Anteon) Hellén, 1935: 7.
opacifrons (Anteon) Hellén, 1935: 7.
rufescens (Gonatopus) Hellén, 1935: 8 (see *Gonatopus helleni* (Raatikainen)).
rufomaculatum (Anteon) Hellén, 1935: 7.

Nomina dubia

fennicus (Anteon) Hellén, 1919: 285.
major (Anteon) Kieffer, 1907: 25.
The original descriptions of the above species are unreliable; their type specimens were not found.

Host records

The following list is based on host records in Fennoscandia and Denmark. Nomenclature follows the monographs by Ossiannilsson (1978, 1981, 1983).

Fulgoromorpha

Cixiidae	*Dryinus niger* Kieffer
Delphacidae	
Criomorphus albomarginatus Cur	*Gonatopus bicolor* (Haliday)
Delphacinus mesomelas (Boh.)	*Gonatopus bicolor* (Haliday)

Dicranotropis hamata (Boh.)	*Gonatopus bicolor* (Haliday)
	Gonatopus dromedarius (A. Costa)
	Gonatopus distinctus Kieffer
Ditropis pteridis (Spinola)	*Gonatopus bicolor* (Haliday)
	Gonatopus distinctus Kieffer
Gravesteiniella boldi (Scott)	*Gonatopus bicolor* (Haliday)
	Gonatopus distinctus Kieffer
Hyledelphax elegantulus (Boh.)	*Gonatopus bicolor* (Haliday)
	Gonatopus distinctus Kieffer
Javesella discolor (Boh.)	*Gonatopus distinctus* Kieffer
Javesella dubia (Kirschbaum)	*Gonatopus formicicolus* (Rich.)
Javesella obscurella (Boh.)	*Gonatopus bicolor* (Haliday)
Javesella pellucida (F.)	*Haplogonatopus oratorius* (West.)
	Gonatopus bicolor (Haliday)
	Gonatopus distinctus Kieffer
	Gonatopus formicicolus (Rich.)
Kelisia sabulicola W. Wagner	*Gonatopus pallidus* (Ceballos)
Kosswigianella exigua (Boh.)	*Gonatopus distinctus* Kieffer
Laodelphax striatellus (Fall.)	*Haplogonatopus oratorius* (West.)
	Gonatopus dromedarius (A. Costa)
Megadelphax sordidulus (Stål)	*Haplogonatopus oratorius* (West.)
	Gonatopus bicolor (Haliday)
	Gonatopus formicicolus (Rich.)
	Gonatopus helleni (Raatikainen)
Megadelphax sp.	*Gonatopus dromedarius* (A. Costa)
Muellerianella fairmairei (Per.)	*Gonatopus formicicolus* (Rich.)
Ribautodelphax angulosus (Rib.)	*Gonatopus distinctus* Kieffer
Ribautodelphax collinus (Boh.)	*Gonatopus bicolor* (Haliday)
Ribautodelphax imitans (Rib.)	*Gonatopus albifrons* Olmi
Ribautodelphax pungens (Rib.)	*Gonatopus bicolor* (Haliday)
	Gonatopus formicicolus (Rich.)
Stiroma bicarinata (H.-S.)	*Gonatopus bicolor* (Haliday)
Unkanodes excisa (Melichar)	*Gonatopus helleni* (Raatikainen)
	Gonatopus bicolor (Haliday)
Xanthodelphax stramineus (Stål)	*Gonatopus formicicolus* (Rich.)
Achilidae	
Cixidia sp.	*Embolemus ruddii* Westwood
Cicadellidae	
Macropsinae	
Oncopsis flavicollis (L.)	*Anteon jurineanum* Latreille
	Anteon brachycerum (Dalman)
Macropsis sp.	*Anteon ephippiger* (Dalman)
	Anteon gaullei Kieffer
	Anteon pubicorne (Dalman)
Idiocerinae	
Idiocerus stigmaticalis Lewis	*Anteon flavicorne* (Dalman)
Idiocerus sp.	*Anteon arcuatum* Kieffer
Populicerus albicans (Kirsch.)	*Anteon flavicorne* (Dalman)
Populicerus confusus (Flor)	*Anteon flavicorne* (Dalman)
Populicerus laminatus (Flor)	*Anteon flavicorne* (Dalman)

Populicerus populi (L.)	*Anteon flavicorne* (Dalman)
Rhytidodus decimusquartus (Sc.)	*Anteon arcuatum* Kieffer
	Anteon flavicorne (Dalman)
Tremulicerus distinguendus (K.)	*Anteon flavicorne* (Dalman)
Iassinae	
Iassus lanio (L.)	*Anteon infectum* (Haliday)
Aphrodinae	
Aphrodes bicincta (Schrank)	*Gonatopus striatus* Kieffer
Typhlocybinae	
Aguriahana germari (Zett.)	*Aphelopus melaleucus* (Dalman)
Alebra albostriella (Fallén)	*Aphelopus atratus* (Dalman)
	Aphelopus serratus Richards
Alebra wahlbergi (Boheman)	*Aphelopus atratus* (Dalman)
Alnetoidia alneti (Dahlbom)	*Aphelopus melaleucus* (Dalman)
	Aphelopus serratus Richards
Chlorita viridula (Fallén)	*Aphelopus camus* Richards
Edwardsiana avellanae (Edw.)	*Aphelopus melaleucus* (Dalman)
Edwardsiana bergmani (Tull.)	*Aphelopus melaleucus* (Dalman)
Edwardsiana crataegi (Douglas)	*Aphelopus melaleucus* (Dalman)
	Aphelopus atratus (Dalman)
	Aphelopus serratus Richards
Edwardsiana flavescens (F.)	*Aphelopus melaleucus* (Dalman)
Edwardsiana geometrica (Schr.)	*Aphelopus melaleucus* (Dalman)
	Aphelopus serratus Richards
Edwardsiana hippocastani (Edw.)	*Aphelopus melaleucus* (Dalman)
	Aphelopus atratus (Dalman)
Edwardsiana lethierryi (Edw.)	*Aphelopus melaleucus* (Dalman)
	Aphelopus atratus (Dalman)
	Aphelopus serratus Richards
Edwardsiana menzbieri Zachv	*Aphelopus melaleucus* (Dalman)
Edwardsiana plebeja (Edw.)	*Aphelopus melaleucus* (Dalman)
Edwardsiana rosae (L.)	*Aphelopus melaleucus* (Dalman)
	Aphelopus atratus (Dalman)
Empoasca decipiens Paoli	*Aphelopus melaleucus* (Dalman)
Empoasca smaragdula (Fallén)	*Aphelopus serratus* Richards
Empoasca solani (Curtis)	*Aphelopus querceus* Olmi
Empoasca vitis (Göthe)	*Aphelopus melaleucus* (Dalman)
	Aphelopus atratus (Dalman)
	Aphelopus serratus Richards
	Aphelopus nigriceps Kieffer
	Aphelopus querceus Olmi
Eupterycyba jucunda (H.-S.)	*Aphelopus serratus* Richards
Eupteryx aurata (L.)	*Aphelopus atratus* (Dalman)
Eupteryx cyclops Matsumura	*Aphelopus atratus* (Dalman)
Eupteryx stachydearum (Hardy)	*Aphelopus atratus* (Dalman)
Eupteryx urticae (F.)	*Aphelopus atratus* (Dalman)
Eurhadina concinna (Edwards)	*Aphelopus nigriceps* Kieffer
Fagocyba carri (Edwards)	*Aphelopus melaleucus* (Dalman)
Fagocyba cruenta (H.-S.)	*Aphelopus melaleucus* (Dalman)
	Aphelopus atratus (Dalman)
	Aphelopus serratus Richards
Fagocyba douglasi (Edwards)	*Aphelopus melaleucus* (Dalman)

Linnavuoriana decempunctata (Fal.)	*Aphelopus melaleucus* (Dalman)
Ossiannilssonola callosa (Th.)	*Aphelopus melaleucus* (Dalman)
Ribautiana ulmi (L.)	*Aphelopus melaleucus* (Dalman)
	Aphelopus atratus (Dalman)
Ribautiana tenerrima (H.-S.)	*Aphelopus atratus* (Dalman)
	Aphelopus serratus Richards
Typhlocyba bifasciata Boheman	*Aphelopus atratus* (Dalman)
Typhlocyba quercus (F.)	*Aphelopus melaleucus* (Dalman)
	Aphelopus atratus (Dalman)
	Aphelopus serratus Richards
Zygina flammigera (Fourcroy)	*Aphelopus melaleucus* (Dalman)
	Aphelopus atratus (Dalman)
Zygina sp.	*Aphelopus serratus* Richards
Deltocephalinae	
Adarrus multinotatus (Boh.)	*Gonatopus lunatus* Klug
	Gonatopus formicarius Ljungh
Adarrus sp.	*Gonatopus spectrum* (Vollenh.)
Arocephalus languidus (Flor)	*Gonatopus clavipes* (Thunb.)
Arocephalus longiceps (Kirsch.)	*Gonatopus clavipes* (Thunb.)
Arocephalus punctum (Flor)	*Anteon pubicorne* (Dalman)
	Gonatopus clavipes (Thunb.)
Arthaldeus pascuellus (Fallén)	*Lonchodryinus ruficornis* (Dal.)
	Anteon pubicorne (Dalman)
	Gonatopus lunatus Klug
	Gonatopus clavipes (Thunb.)
Arthaldeus striifrons (Kirsch.)	*Gonatopus clavipes* (Thunb.)
Cicadula sp.	*Gonatopus clavipes* (Thunb.)
Conosanus obsoletus (Kir.)	*Lonchodryinus ruficornis* (Dal.)
Elymana sulphurella (Zett.)	*Lonchodryinus ruficornis* (Dal.)
	Gonatopus clavipes (Thunb.)
Errastunus ocellaris (Fallén)	*Lonchodryinus ruficornis* (Dal.)
	Gonatopus clavipes (Thunb.)
Errastunus sp.	*Gonatopus lunatus* Klug
Euscelis incisus (Kirsch.)	*Lonchodryinus ruficornis* (Dal.)
	Anteon pubicorne (Dalman)
	Gonatopus lunatus Klug
	Gonatopus clavipes (Thunb.)
Graphocraerus ventralis (Fall.)	*Anteon tripartitum* Kieffer
Jassargus distinguendus (Flor)	*Lonchodryinus ruficornis* (Dal.)
	Gonatopus clavipes (Thunb.)
Jassargus flori (Fieber)	*Lonchodryinus ruficornis* (Dal.)
	Gonatopus clavipes (Thunb.)
Jassargus sp.	*Gonatopus spectrum* (Vollenh.)
Macrosteles frontalis (Scott)	*Anteon fulviventre* (Haliday)
Macrosteles laevis (Ribaut)	*Lonchodryinus ruficornis* (Dal.)
	Anteon ephippiger (Dalman)
	Gonatopus pedestris Dalman
Macrosteles quadripunctulatus (Kir.)	*Gonatopus pedestris* Dalman
Macrosteles sexnotatus (Fall.)	*Anteon ephippiger* (Dalman)
	Anteon pubicorne (Dalman)
Macrosteles viridigriseus (Ed.)	*Anteon pubicorne* (Dalman)
Macrosteles sp.	*Gonatopus lunatus* Klug
Mocuellus collinus (Boh.)	*Gonatopus distinguendus* Kieffer

Opsius stactogalus Fieber	*Anteon ephippiger* (Dalman)
	Anteon pubicorne (Dalman)
Psammotettix alienus (Dahlb.)	*Gonatopus lunatus* Klug
	Gonatopus formicarius Ljungh
	Gonatopus clavipes (Thunb.)
Psammotettix cephalotes (H.-S.)	*Lonchodryinus ruficornis* (Dal.)
	Gonatopus distinguendus Kieffer
	Gonatopus formicarius Ljungh
	Gonatopus clavipes (Thunb.)
Psammotettix confinis (Dahlb.)	*Lonchodryinus ruficornis* (Dal.)
	Anteon pubicorne (Dalman)
	Gonatopus distinguendus Kieffer
	Gonatopus formicarius Ljungh
	Gonatopus clavipes (Thunb.)
Psammotettix nodosus (Ribaut)	*Lonchodryinus ruficornis* (Dal.)
	Anteon pubicorne (Dalman)
	Gonatopus clavipes (Thunb.)
Psammotettix putoni (Then)	*Gonatopus clavipes* (Thunb.)
Rhopalopyx sp.	*Gonatopus spectrum* (Vollenh.)
Sorhoanus assimilis (Fall.)	*Gonatopus clavipes* (Thunb.)
Sorhoanus xanthoneurus (Fieb.)	*Gonatopus clavipes* (Thunb.)
Streptanus sordidus (Zett.)	*Lonchodryinus ruficornis* (Dal.)
	Anteon pubicorne (Dalman)
	Gonatopus clavipes (Thunb.)
Thamnotettix confinis (Zett.)	*Anteon tripartitum* Kieffer
	Gonatopus clavipes (Thunb.)
Turrutus socialis (Flor)	*Gonatopus clavipes* (Thunb.)
Verdanus abdominalis (F.)	*Gonatopus formicarius* Ljungh
	Gonatopus clavipes (Thunb.)

Colour plates 1–38

Plate N. 1. Nymph of *Psammotettix* sp. parasitized by *Gonatopus clavipes* (Thunberg).

Plate N. 2. Cocoon of Dryininae.

Plate N. 3. Mature larva of *Anteon ephippiger* (Dalman) emerging from a parasitized nymph of *Opsius* sp.

Plate N. 4. Mature larva of *Anteon ephippiger* (Dalman) just after emergence from a nymph of *Opsius* sp. and searching for a pupation site.

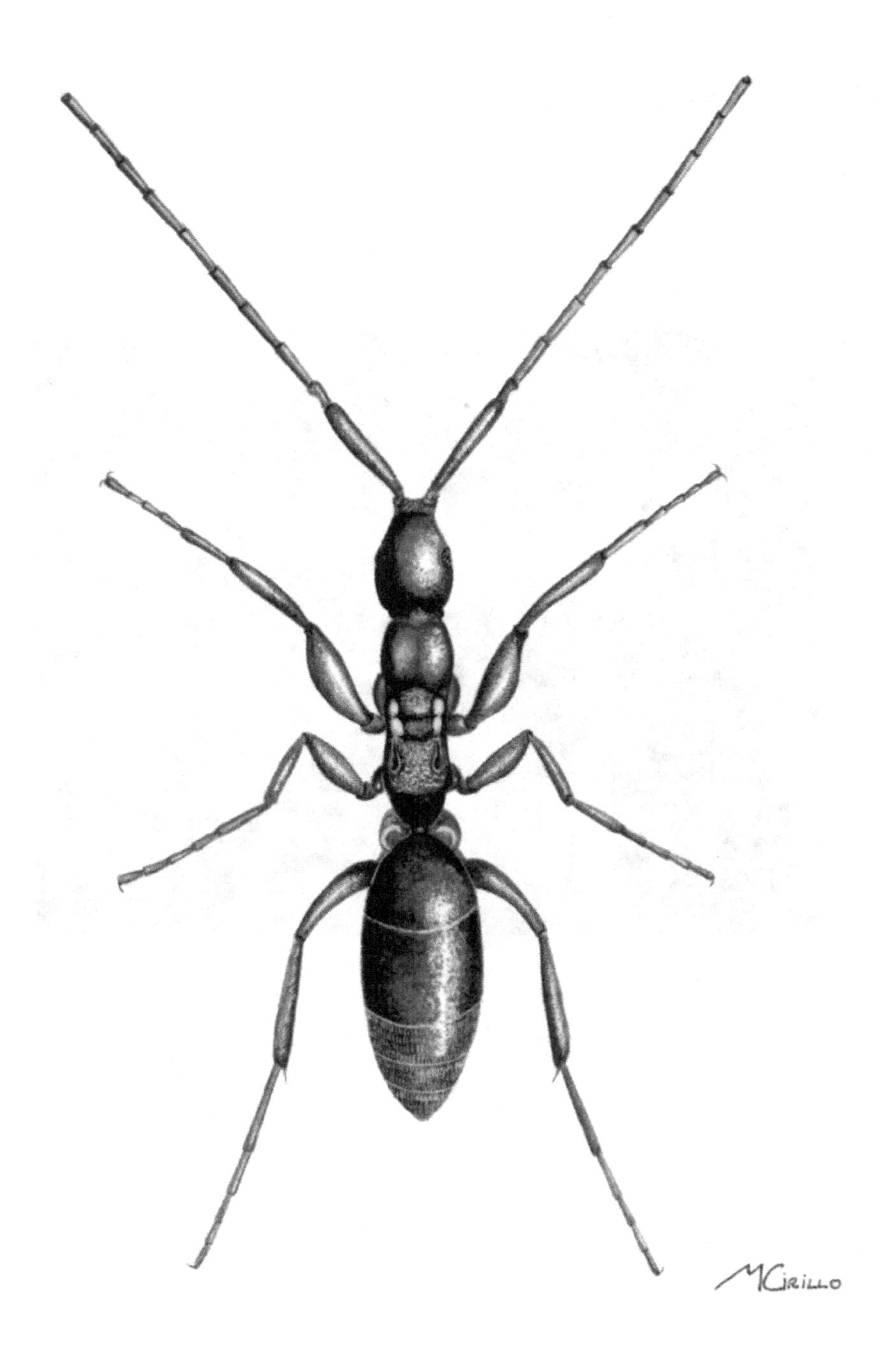

Plate N. 5. Female of *Embolemus ruddii* Westwood.

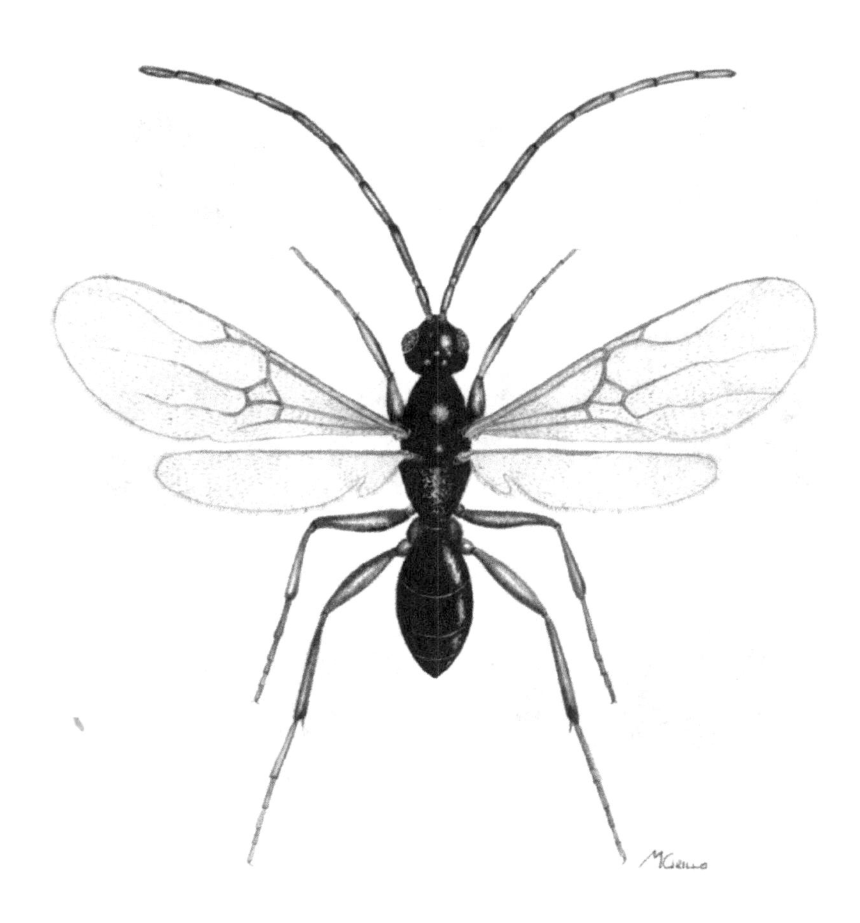

Plate N. 6. Male of *Embolemus ruddii* Westwood.

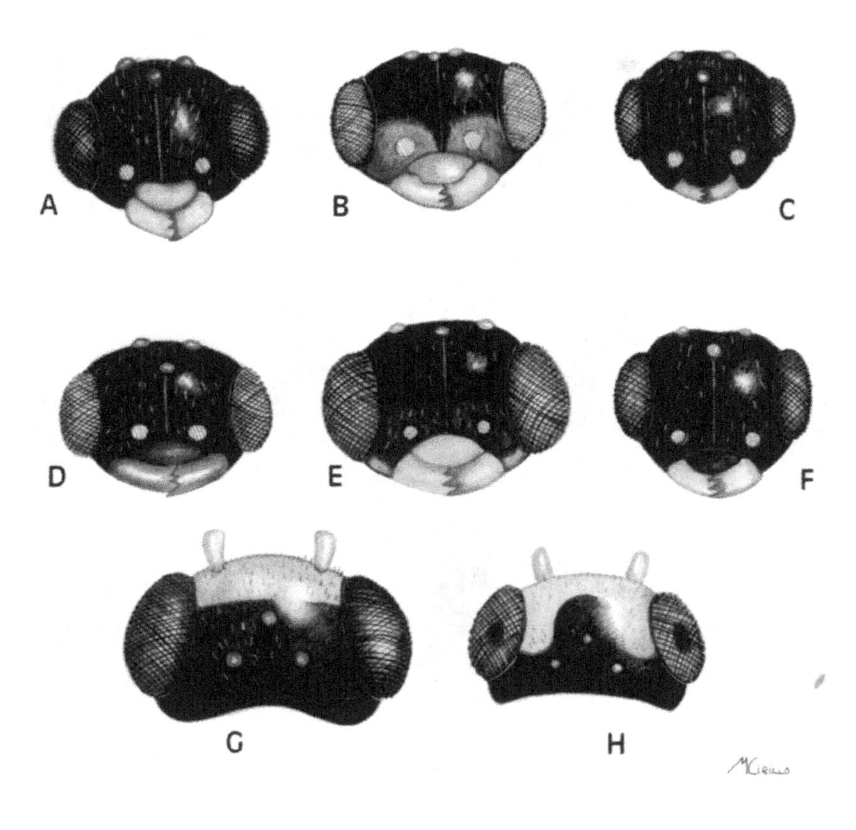

Plate N. 7. Heads of *Aphelopus* spp. : A: *serratus* (male); B: *melaleucus* (male); C: *atratus* (male); D: *camus* (male); E: *querceus* (male); F: *nigriceps* (male); G: *querceus* (female); H: *melaleucus* (female).

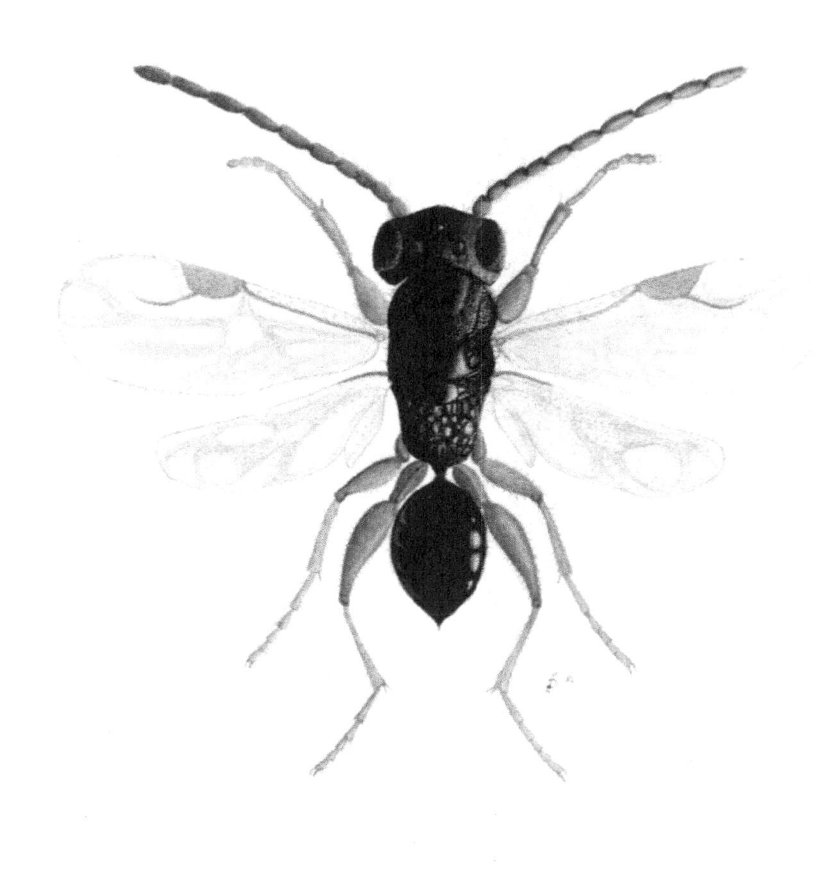

Plate N. 8. Male of *Aphelopus melaleucus* (Dalman).

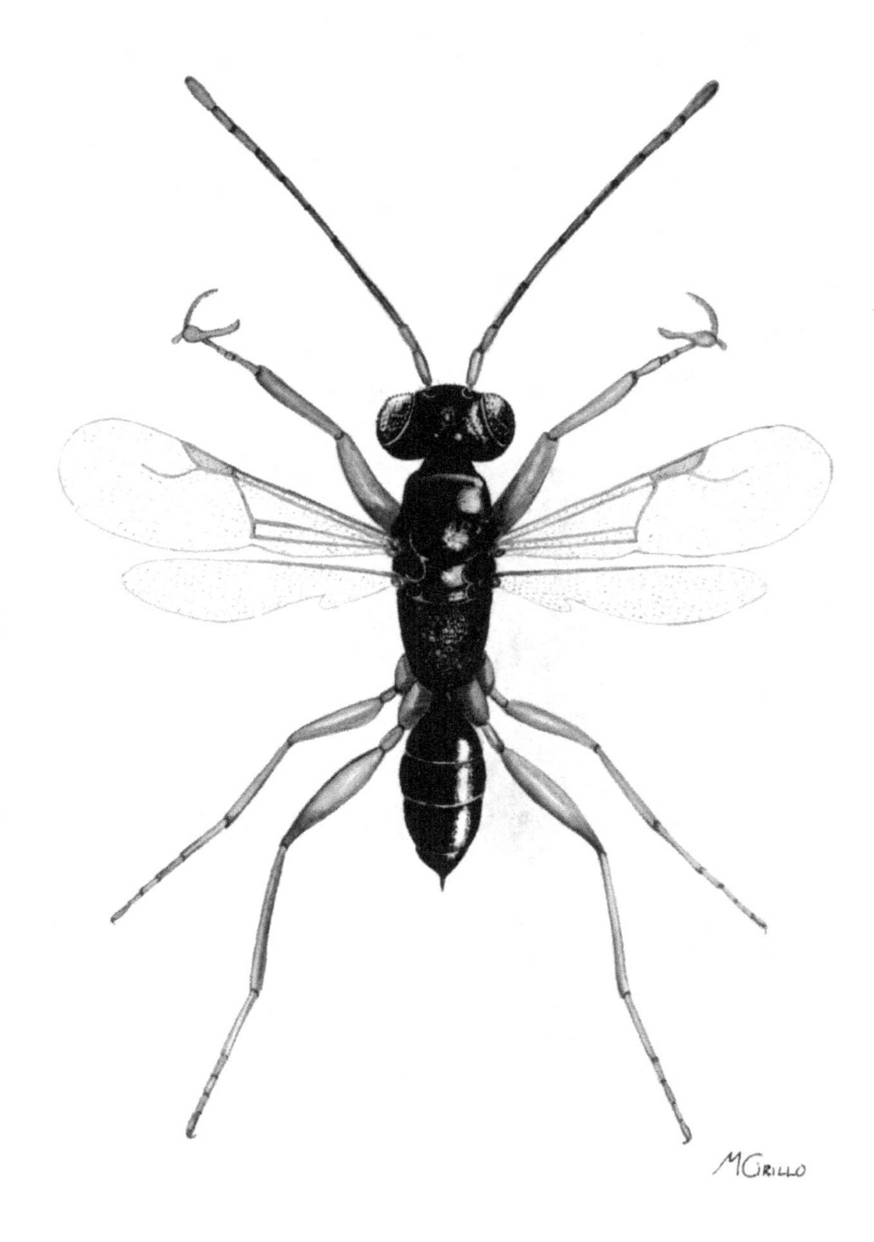

Plate N. 9. Female of *Lonchodryinus ruficornis* (Dalman).

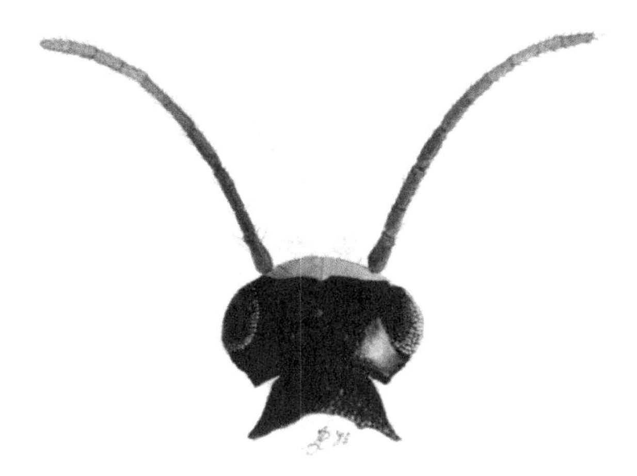

Plate N. 10. Head of female of *Lonchodryinus ruficornis* (Dalman): variety with partly yellow frons.

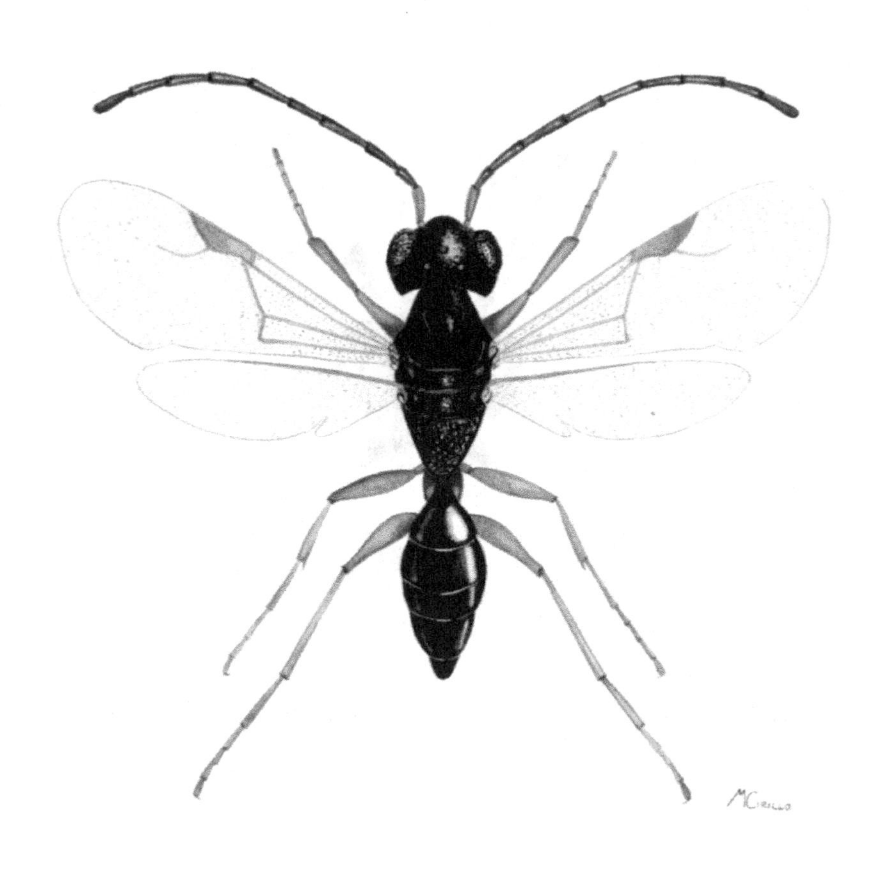

Plate N. 11. Male of *Lonchodryinus ruficornis* (Dalman).

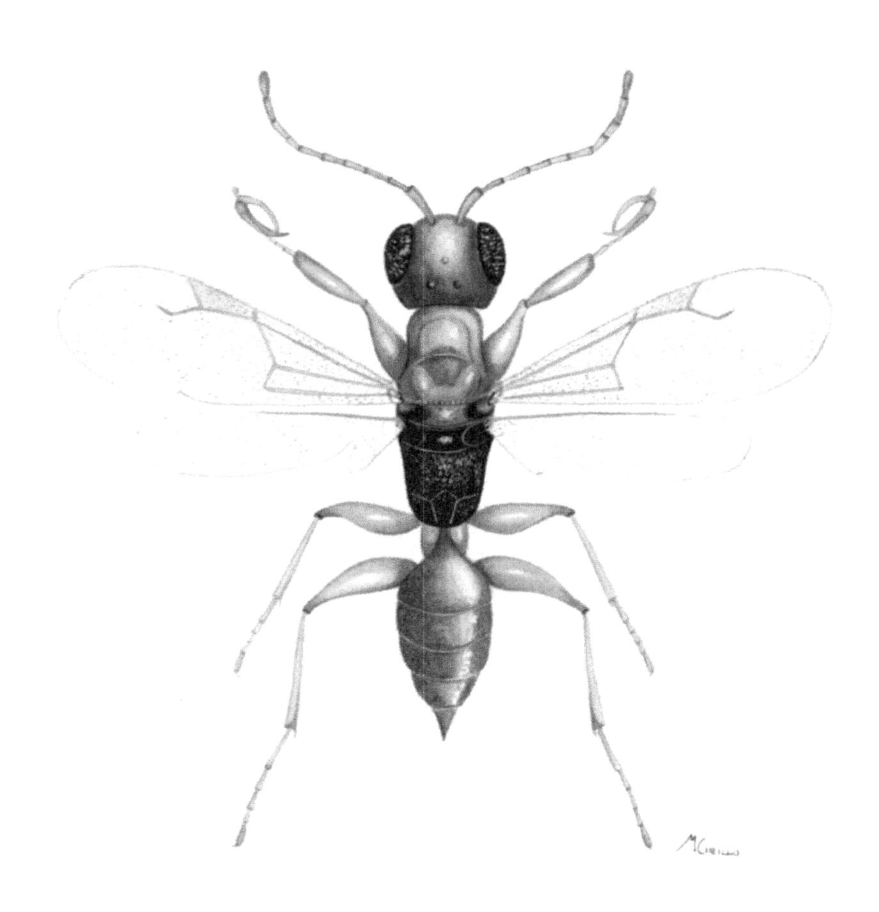

Plate N. 12. Female of *Anteon ephippiger* (Dalman).

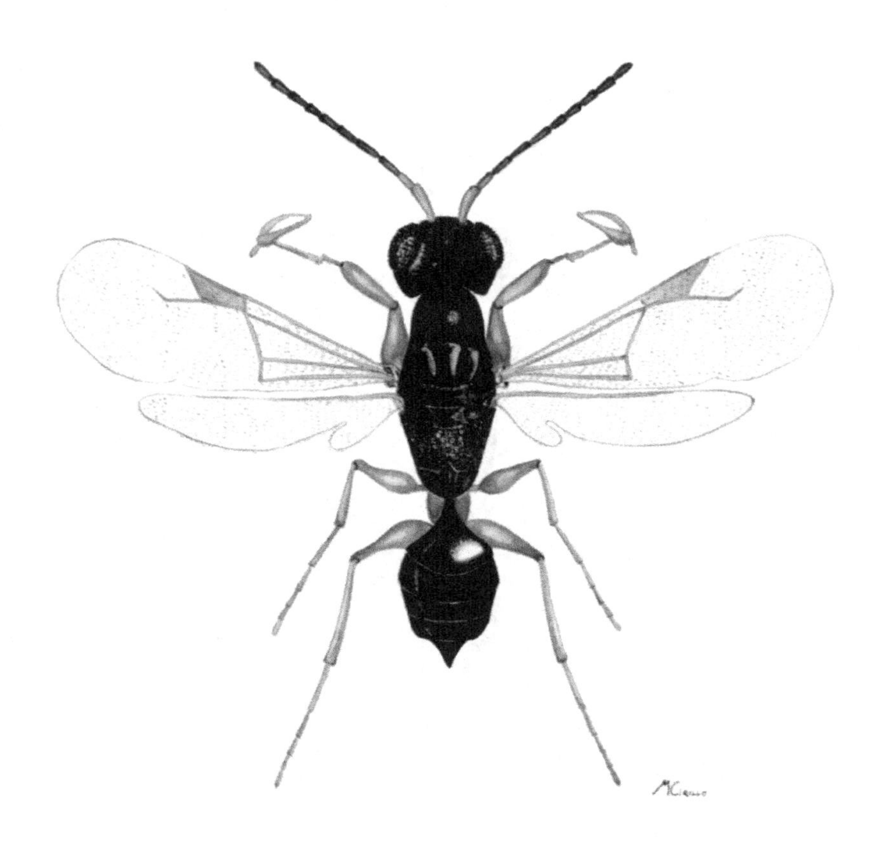

Plate N. 13. Female of *Anteon pubicorne* (Dalman).

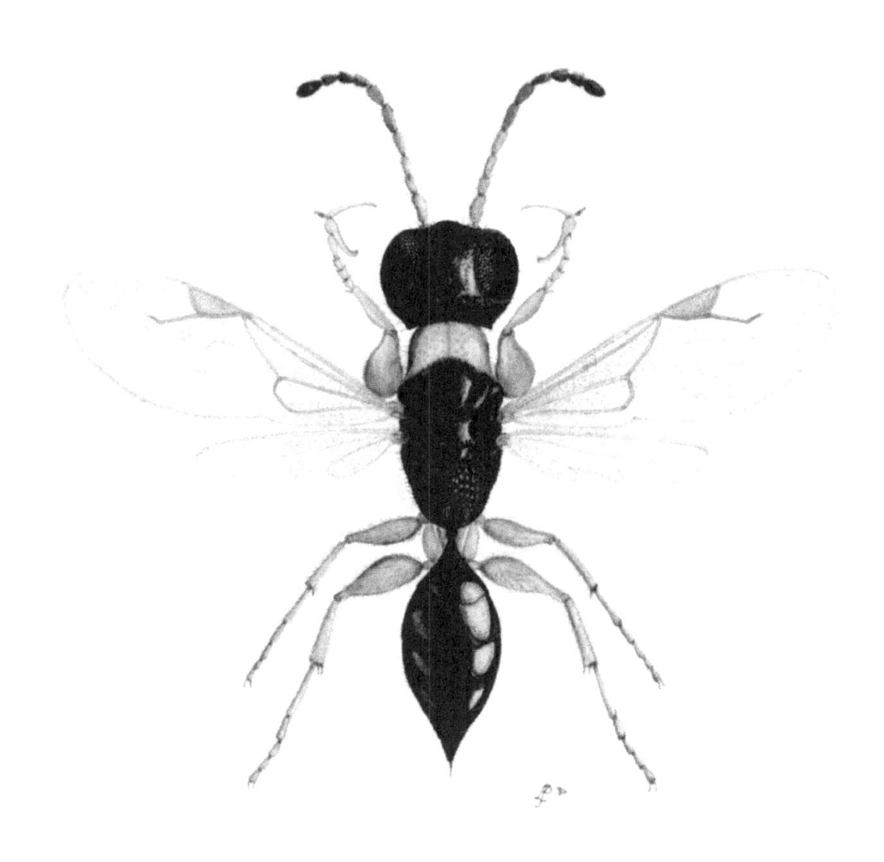

Plate N. 14. Female of *Anteon gaullei* Kieffer.

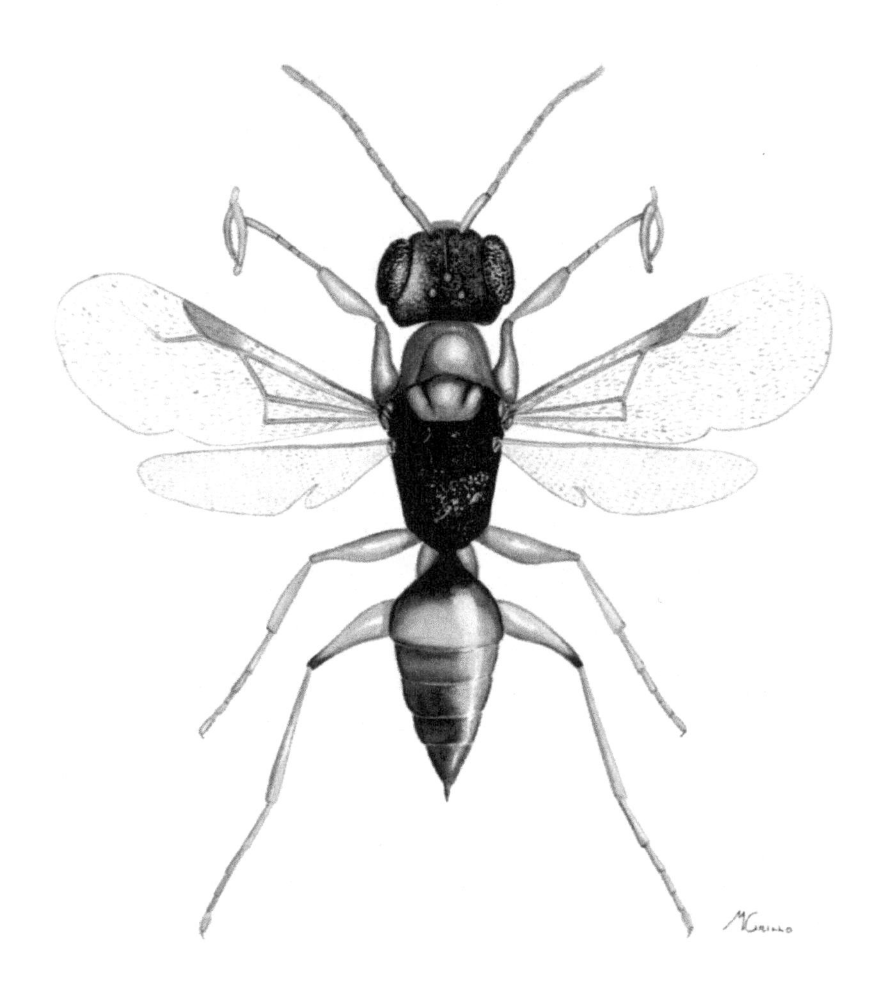

Plate N. 15. Female of *Anteon fulviventre* (Haliday).

Plate N. 16. Male of *Anteon ephippiger* (Dalman).

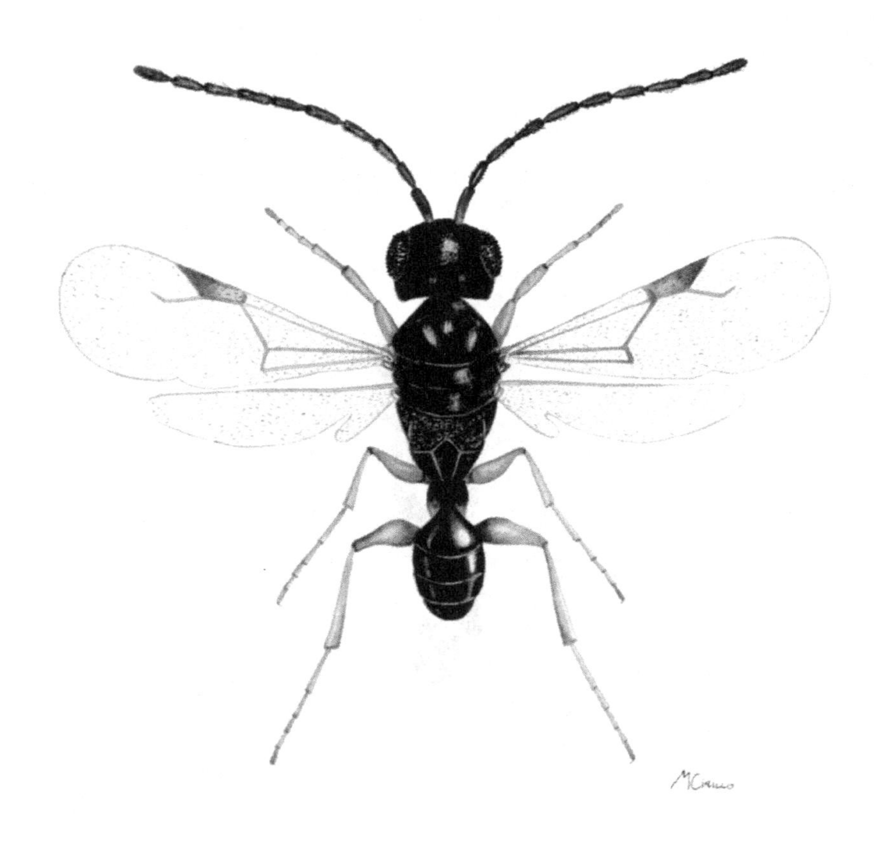

Plate N. 17. Male of *Anteon gaullei* Kieffer.

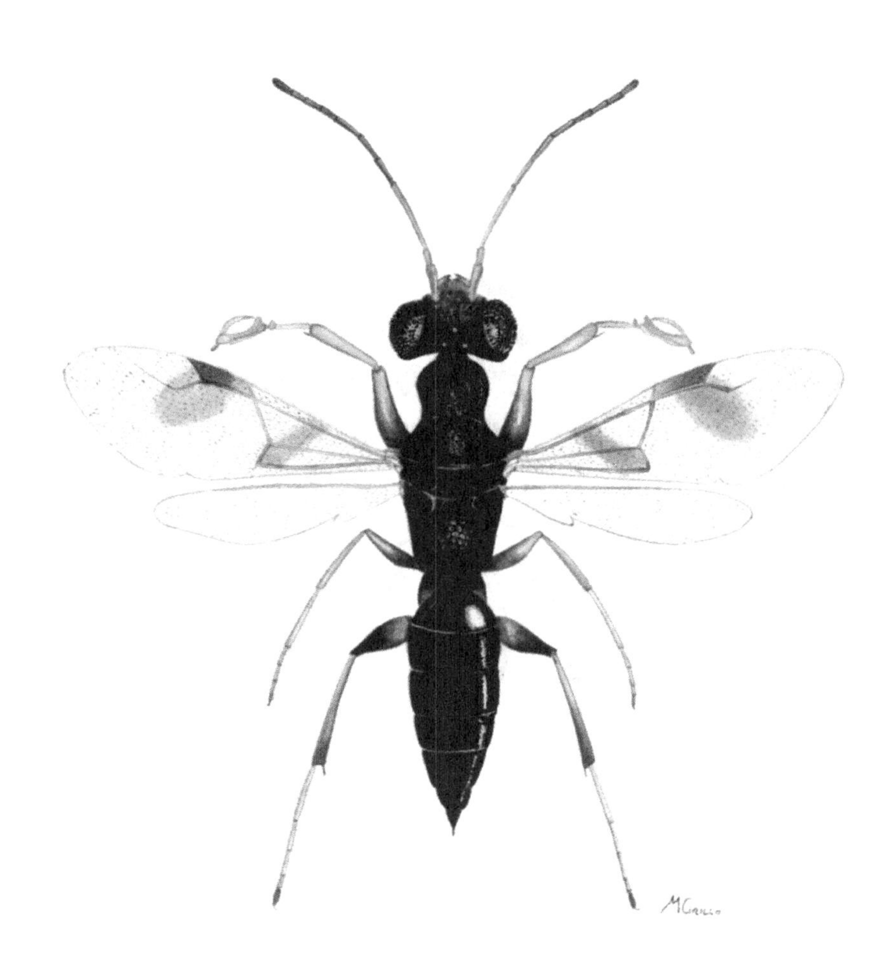

Plate N. 18. Female of *Dryinus niger* Kieffer.

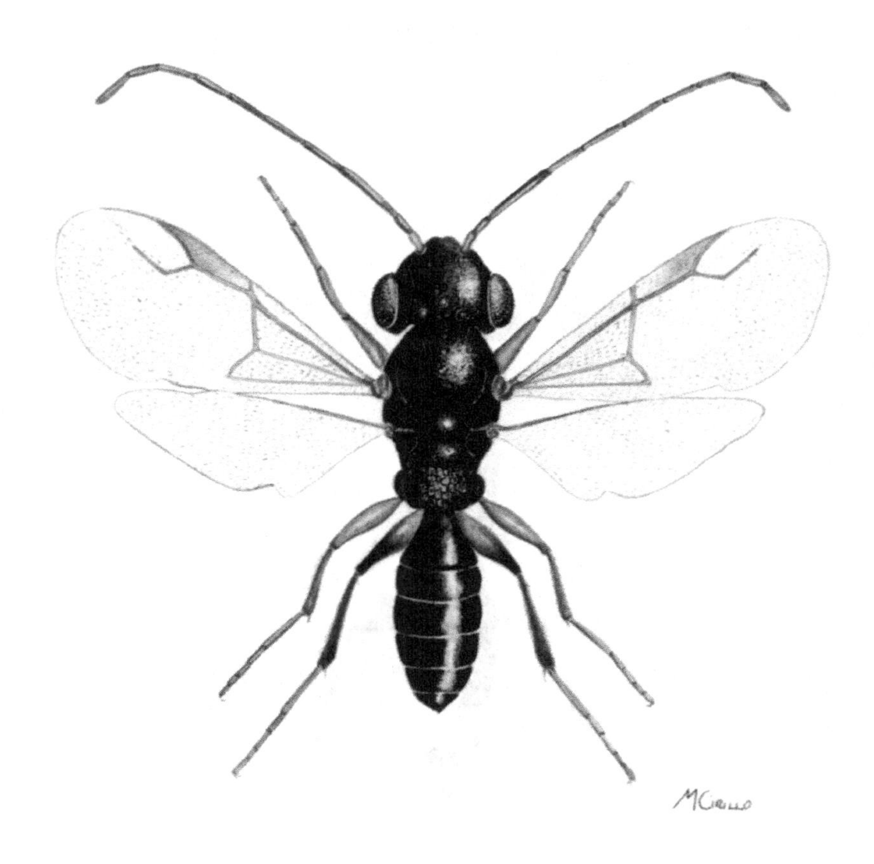

Plate N. 19. Male of *Dryinus niger* Kieffer.

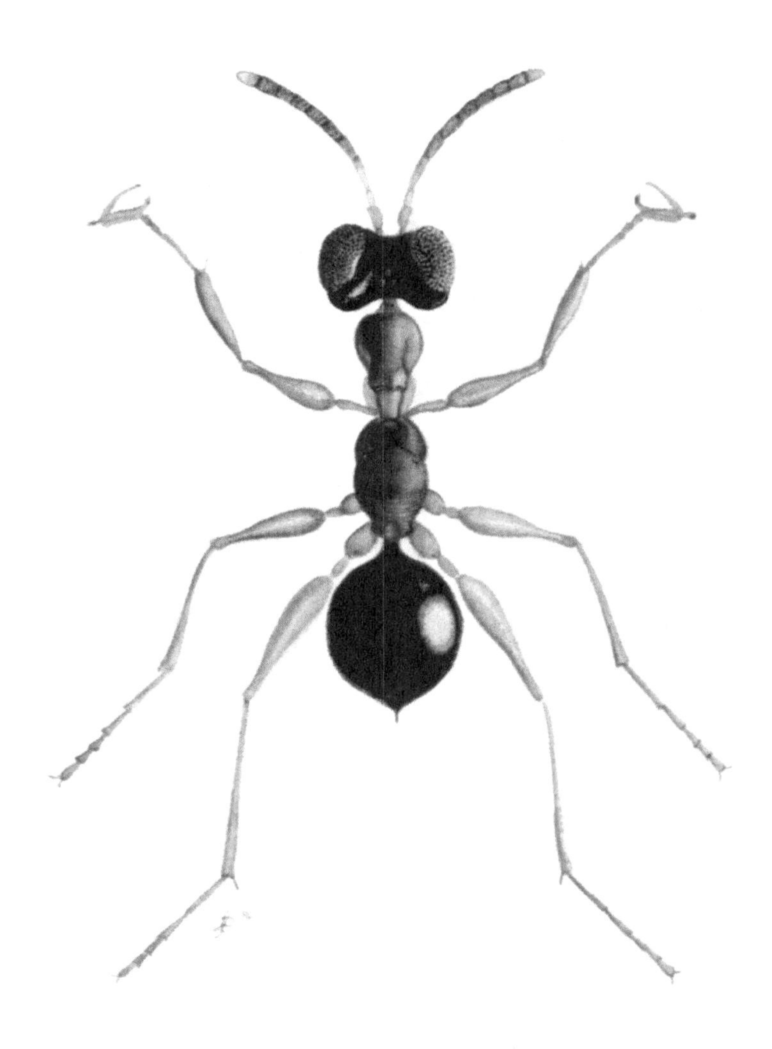

Plate N. 20. Female of *Haplogonatopus oratorius* (Westwood).

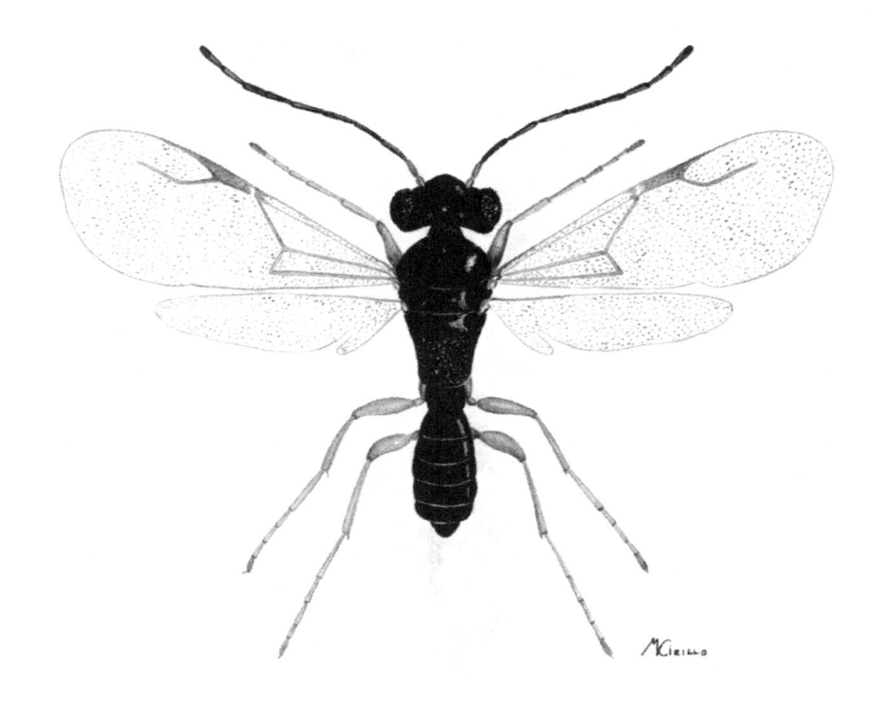

Plate N. 21. Male of *Haplogonatopus oratorius* (Westwood).

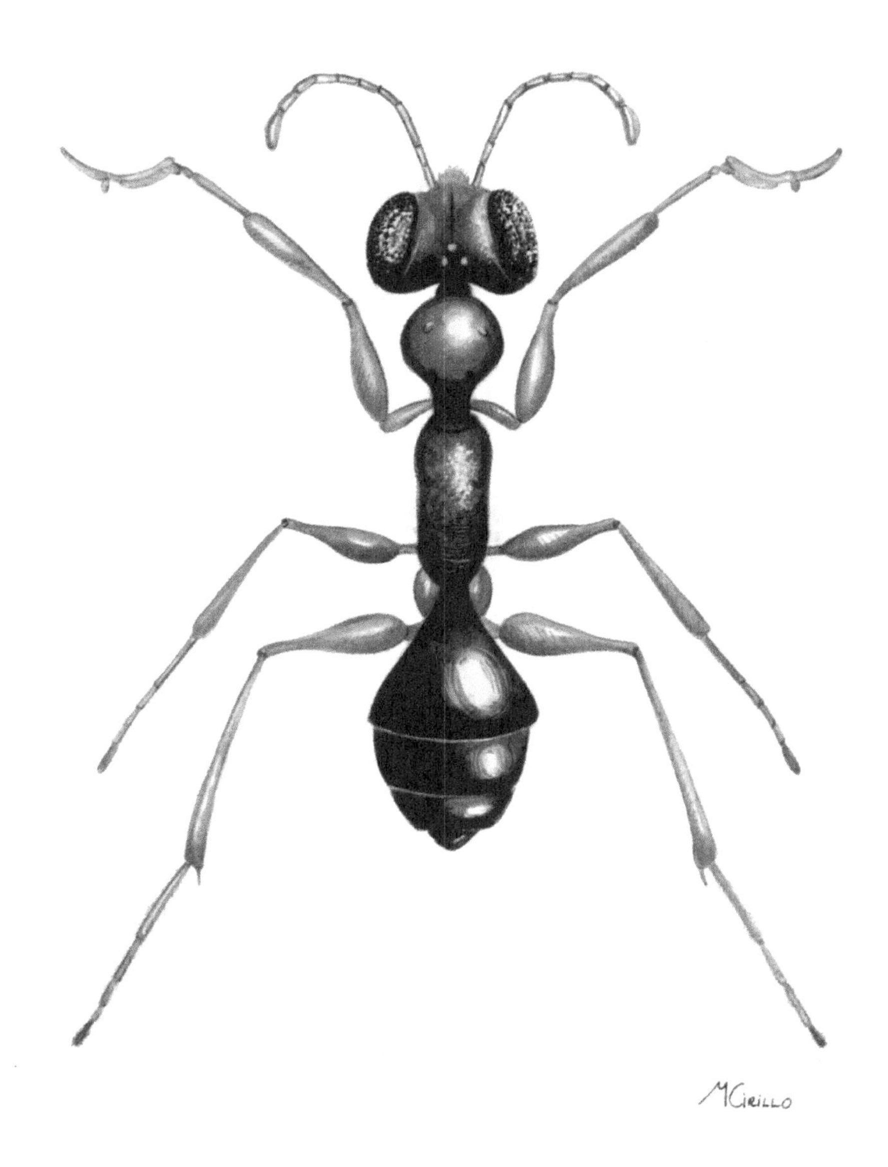

Plate N. 22. Female of *Gonatopus helleni* (Raatikainen).

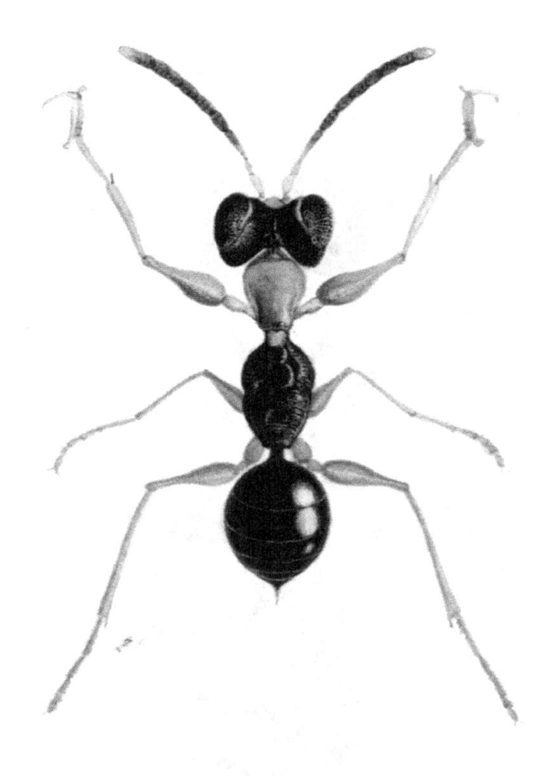

Plate N. 23. Female of *Gonatopus bicolor* (Haliday).

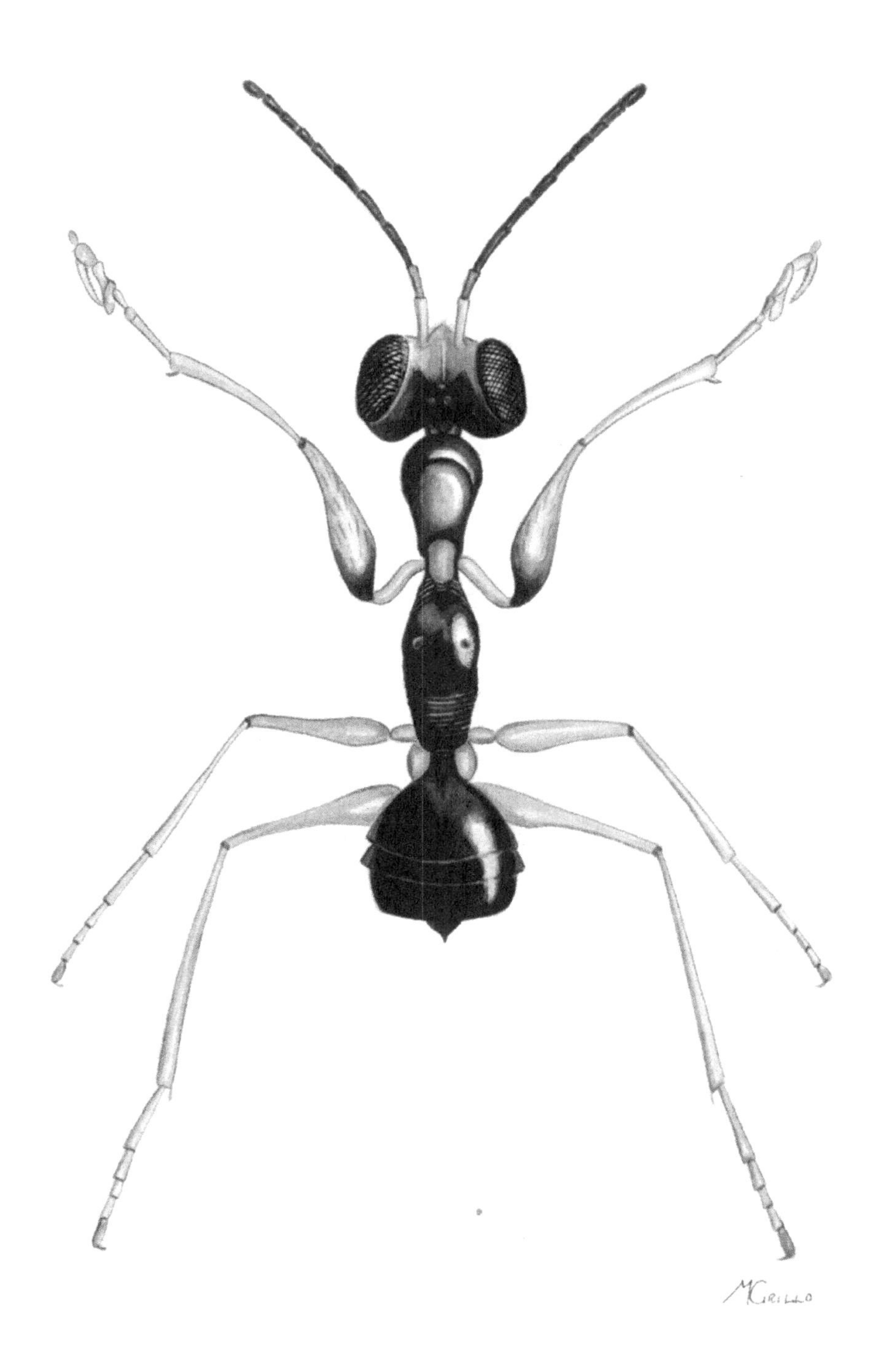

Plate N. 24. Female of *Gonatopus pallidus* (Ceballos).

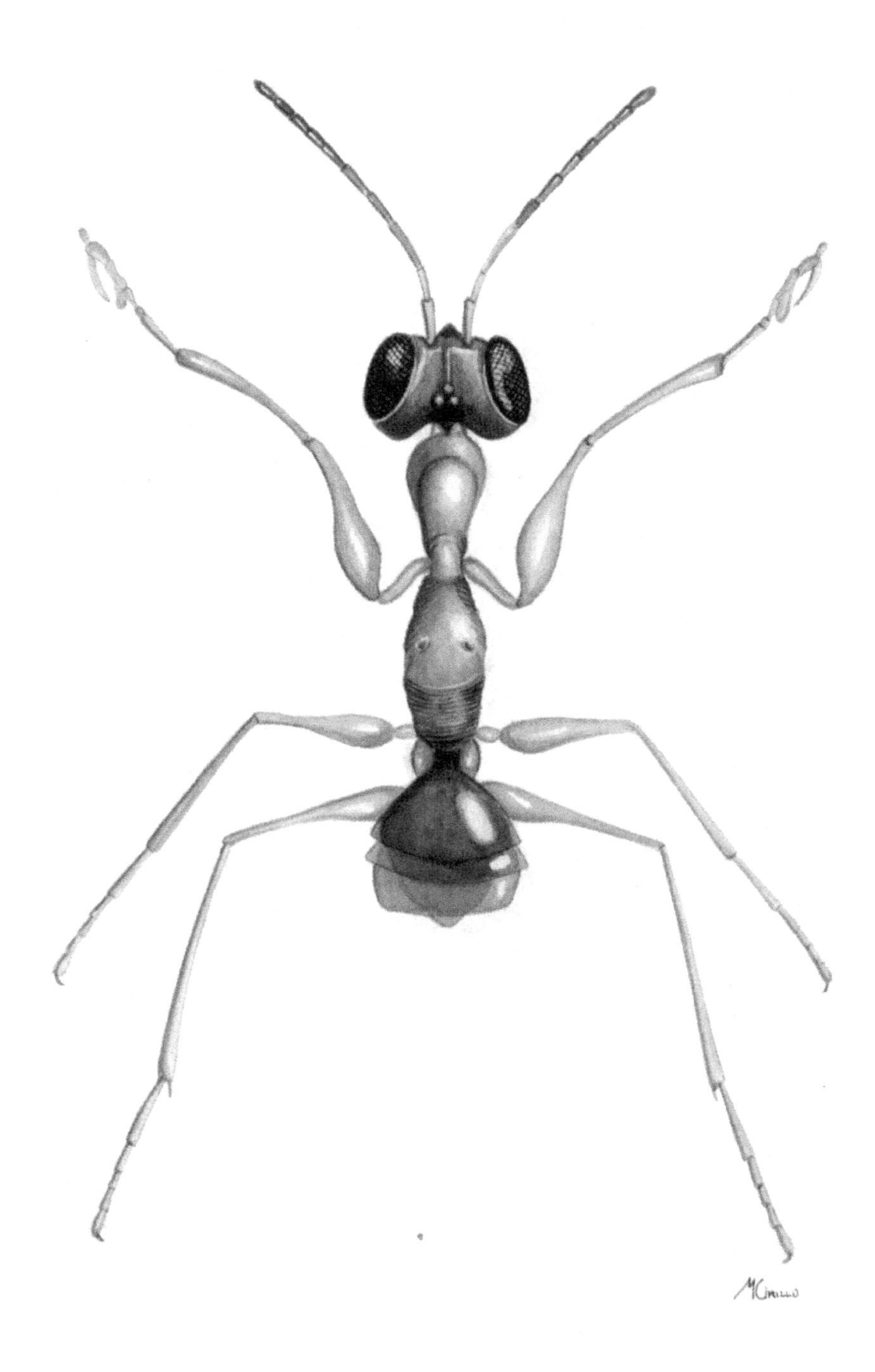

Plate N. 25. Female of *Gonatopus formicicolus* (Richards)

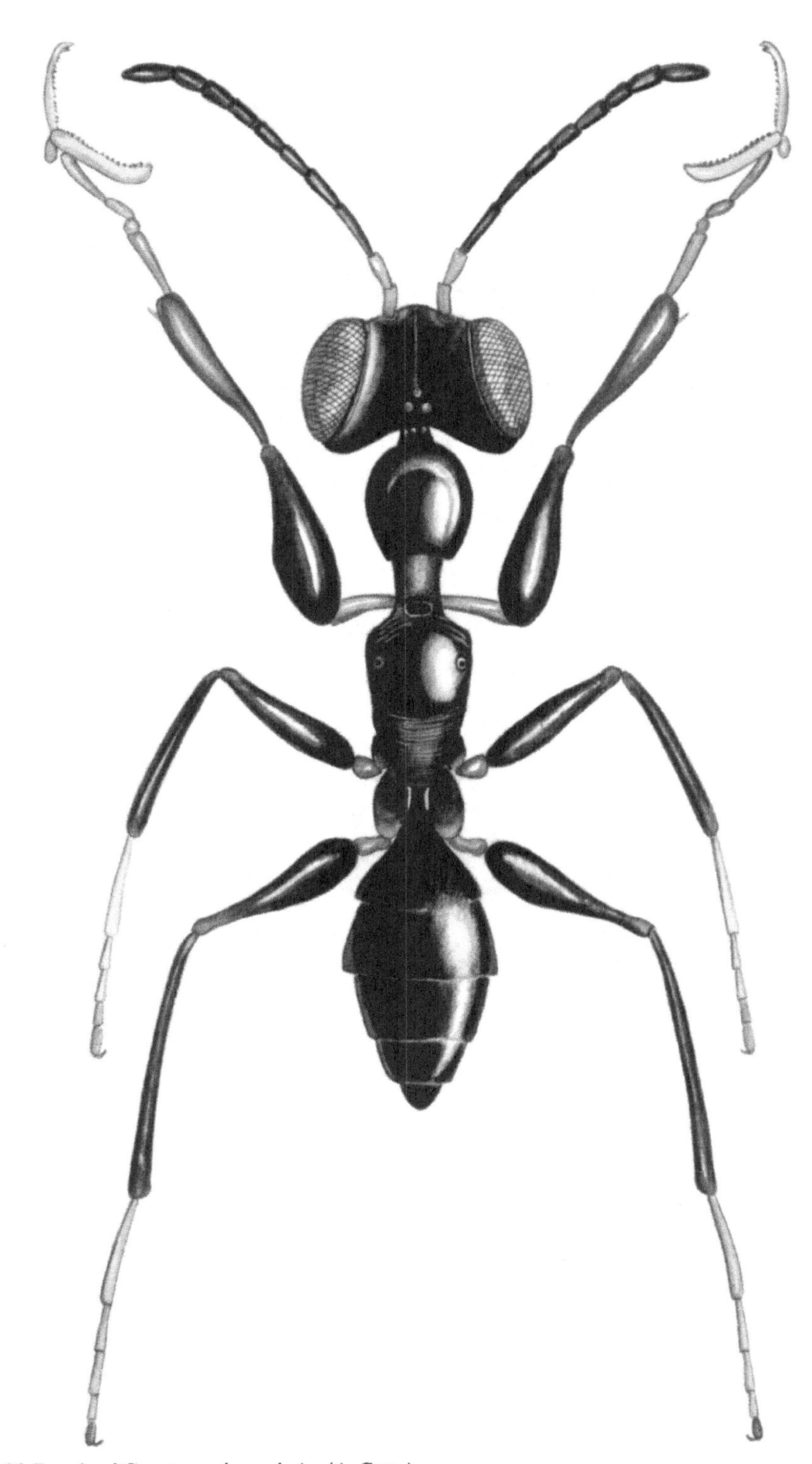

Plate N. 26. Female of *Gonatopus dromedarius* (A. Costa).

Plate N. 27. Female of *Gonatopus distinctus* Kieffer.

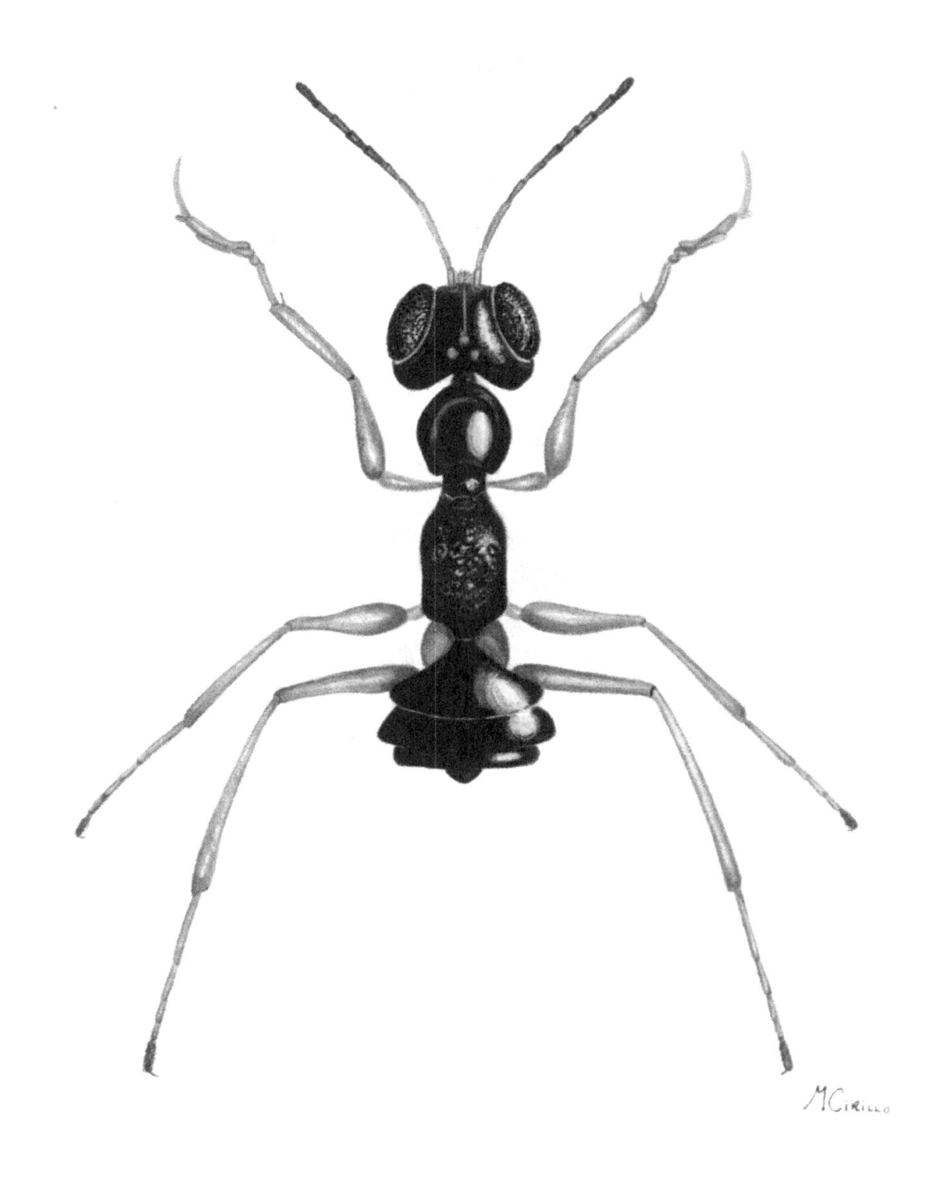

Plate N. 28. Female of *Gonatopus pedestris* Dalman.

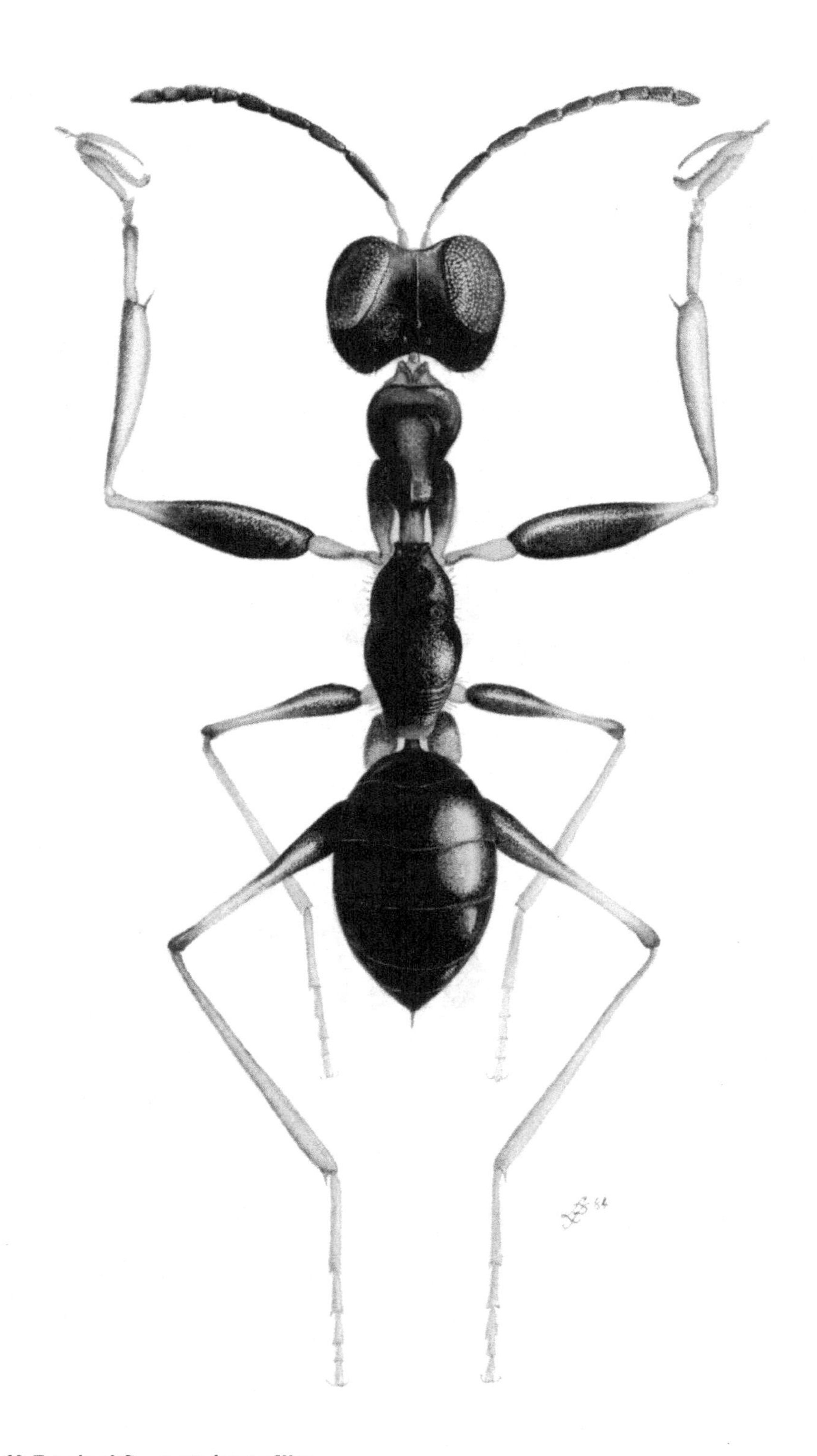

Plate N. 29. Female of *Gonatopus lunatus* Klug.

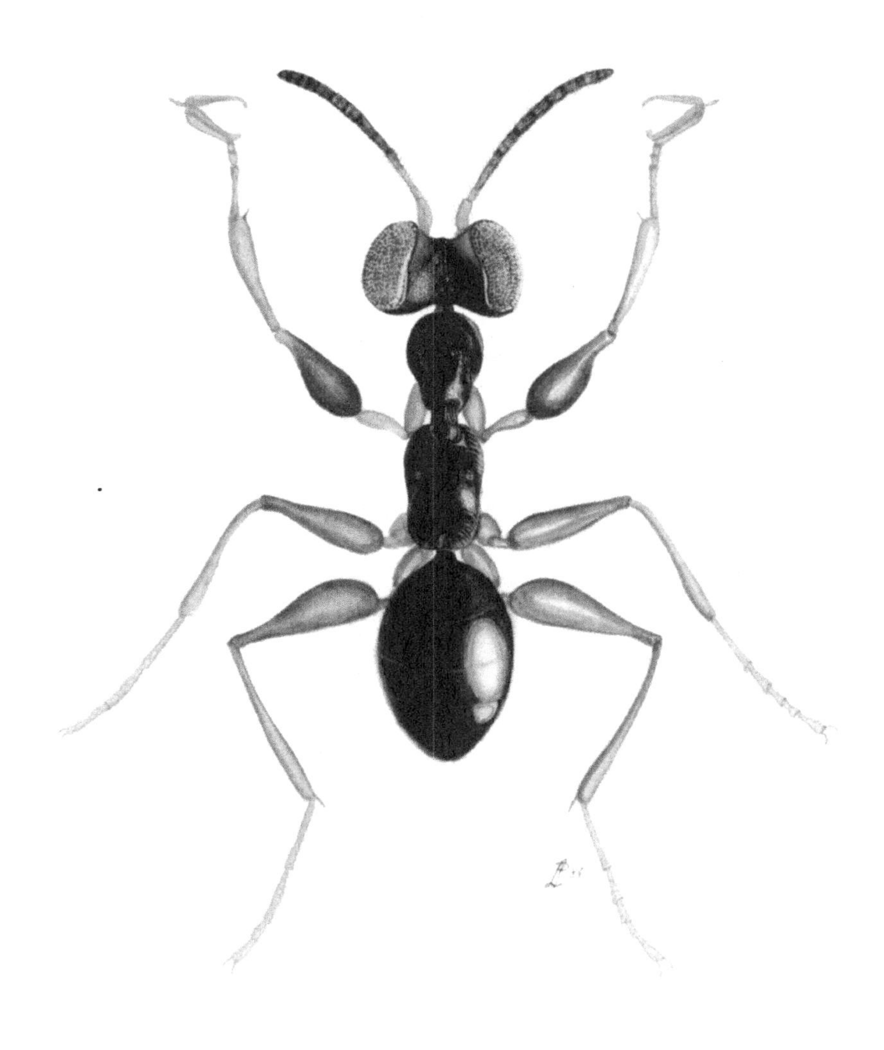

Plate N. 30. Female of *Gonatopus striatus* Kieffer.

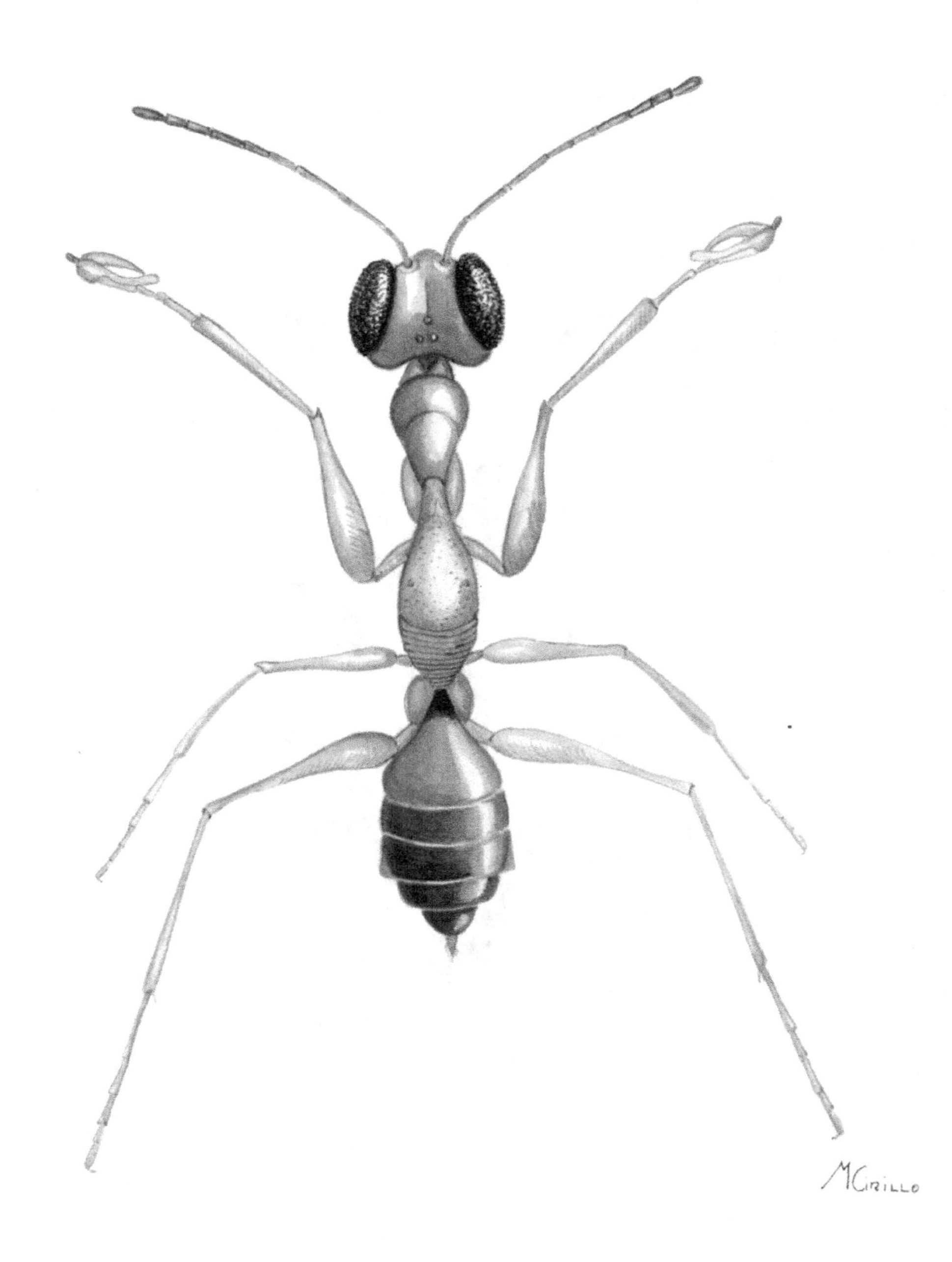

Plate N. 31. Female of *Gonatopus spectrum* (Vollenhoven).

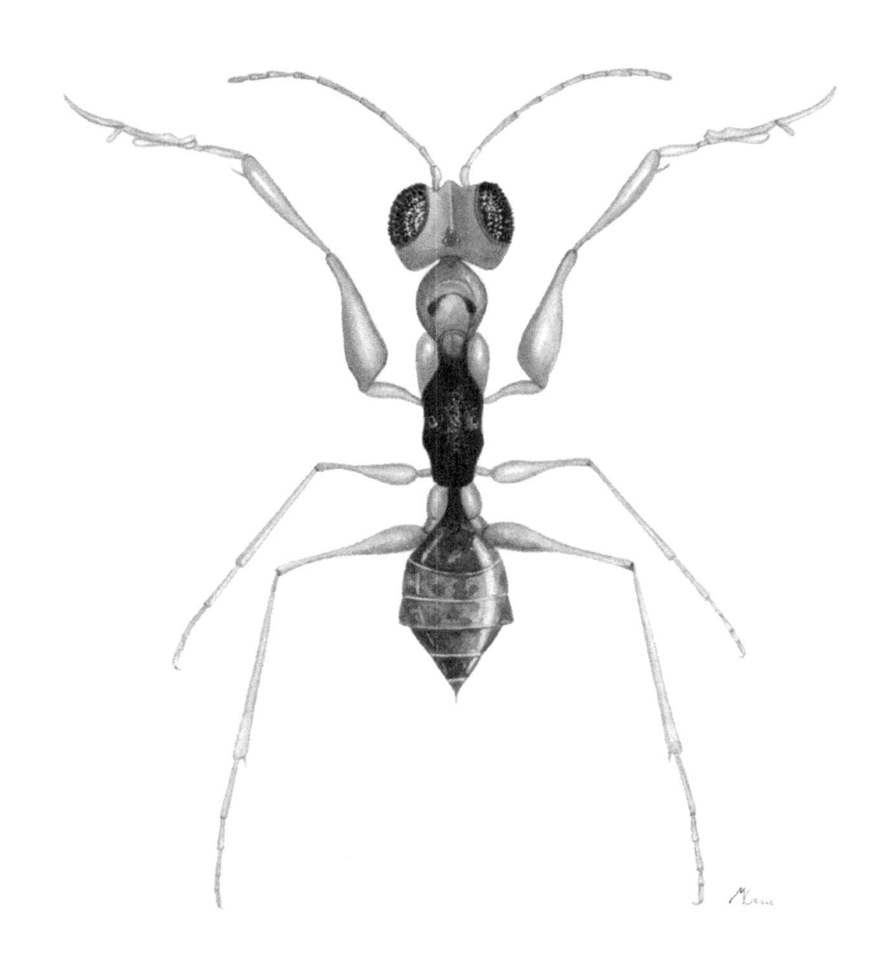

Plate N. 32. Female of *Gonatopus distinguendus* Kieffer: coloured variety.

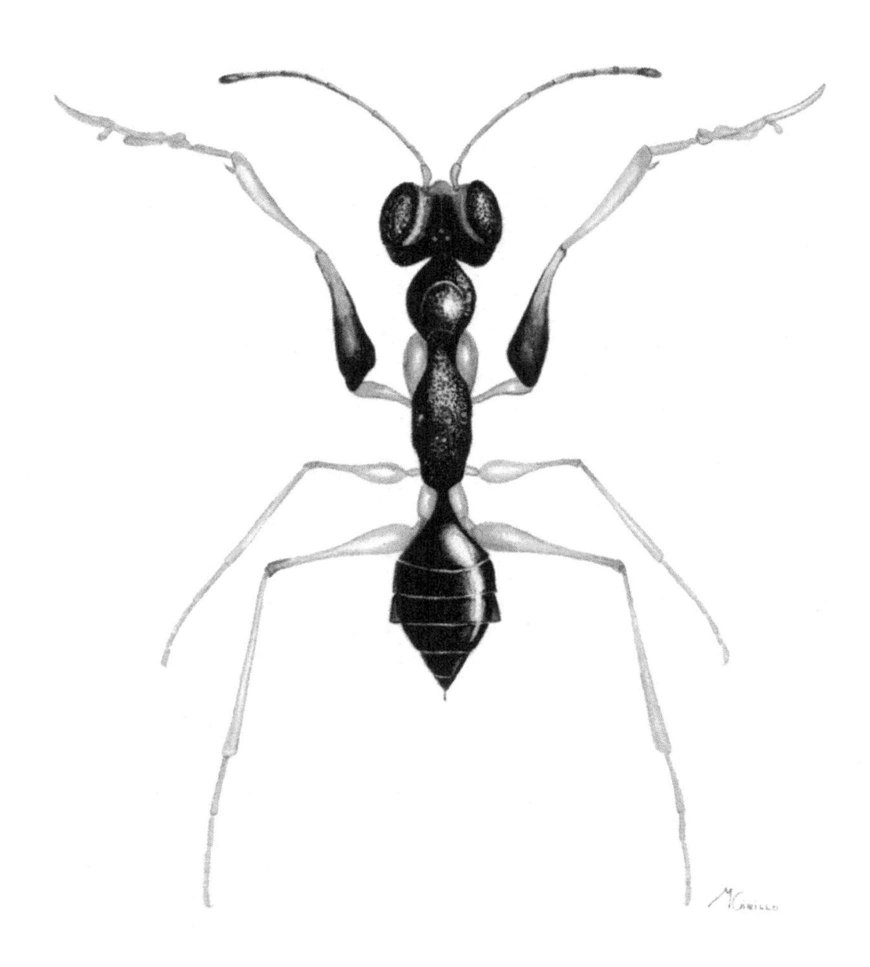

Plate N. 33. Female of *Gonatopus distinguendus* Kieffer: black variety.

Plate N. 34. Female of *Gonatopus formicarius* Ljungh.

Plate N. 35. Female of *Gonatopus clavipes* (Thunberg).

Plate N. 36. Male of *Gonatopus clavipes* (Thunberg).

Plate N. 37. Female of *Gonatopus clavipes* (Thunberg) approaching a nymph of *Psammotettix* sp.

Plate N. 38. Female of *Gonatopus clavipes* (Thunberg) supporting herself on mid and hind tarsi and firmly holding a nymph of *Psammotettix* sp. in a close grip with pincers and apex of gaster.

Catalogue

		NORWAY																			FINLAND																				RUSSIA		
		Ø + AK	HE (s + n)	O (s + n)	B (ø +v)	VE	TE (y + i)	AA (y + i)	VA (y + i)	R (y + i)	HO (y + i)	SF (y + i)	MR (y + i)	ST (y + i)	NT (y + i)	Ns (y + i)	Nn (ø + v)	TR (y + i)	F (v + i)	F (n + ø)	Al	Ab	N	Ka	St	Ta	Sa	Oa	Tb	Sb	Kb	Om	Ok	Ob S	Ob N	Ks	Lk W	Lk E	Le	Li	Ib	Kr	Lr
Embolemus ruddii Westwood	1				●		●				●											●	●					●													●		
Aphelopus melaleucus (Dalman)	2	●			●	●	●	●			●										●	●	●			●	●		●				●			●				●	●	●	
A. atratus (Dalman)	3	●			●	●	●														●	●	●		●		●			●	●										●		
A. serratus Richards	4	●		●	●	●		●													●	●	●			●																	
A. camus Richards	5			●	●	●																					●																
A. nigriceps Kieffer	6						●														●	●	●				●																
A. querceus Olmi	7	●			●	●																●					●																
Lonchodryinus ruficornis (Dalman)	8	●		●	●	●		●							●	●			●		●	●	●		●	●	●		●	●	●		●		●	●	●		●	●	●	●	●
Anteon jurineanum Latreille	9				●		●										●				●	●	●	●	●	●	●	●	●		●		●			●	●			●	●	●	
A. brachycerum (Dalman)	10	●					●				●										●	●	●			●	●		●													●	
A. arcuatum Kieffer	11	●																					●		●		●	●	●		●											●	
A. flavicorne (Dalman)	12	●					●															●	●		●	●	●			●											●		
A. ephippiger (Dalman)	13				●						●											●	●		●	●	●		●			●									●		
A. pubicorne (Dalman)	14	●		●	●	●		●			●					●					●	●	●		●	●	●		●	●		●	●		●				●	●	●	●	●
A. infectum (Haliday)	15	●																					●																				
A. exiguum (Haupt)	16	●			●	●	●	●														●	●																	●			
A. tripartitum Kieffer	17	●			●			●													●	●	●		●		●														●	●	
A. gaullei Kieffer	18	●			●	●		●													●	●	●			●	●	●	●												●		
A. fulviventre (Haliday)	19	●	●		●			●													●	●	●		●	●	●		●												●		
Dryinus niger Kieffer	20						●																																				
Haplogonatopus oratorius (Westwood)	21																																								●		
Gonatopus helleni (Raatikainen)	22			●	●																	●	●			●	●	●	●														
G. bicolor (Haliday)	23					●				●											●							●													●		
G. pallidus (Ceballos)	24																																										
G. formicicolus (Richards)	25																						●																				
G. dromedarius (A. Costa)	26																																										
G. distinctus Kieffer	27				●					●											●		●			●	●	●													●	●	
G. pedestris Dalman	28	●																			●	●	●		●		●							●							●	●	
G. lunatus Klug	29				●	●																●	●		●	●	●														●		
G. striatus Kieffer	30	●			●																	●	●																				
G. spectrum (Van Vollenhoven)	31																																										
G. distinguendus Kieffer	32																				●	●	●																		●		
G. formicarius Ljungh	33																					●				●															●		
G. clavipes (Thunberg)	34	●			●	●	●													●	●	●	●		●	●	●	●	●				●	●		●					●		●
G. albifrons Olmi	35																																										

				DENMARK											SWEDEN																													
		Germany	G. Britain	SJ	EJ	WJ	NWJ	NEJ	F	LFM	SZ	NWZ	NEZ	B	Sk.	Bl.	Hall.	Sm.	Öl.	Gtl.	G. Sand.	Ög.	Vg.	Boh.	Dlsl.	Nrk.	Sdm.	Upl.	Vstm.	Vrm.	Dlr.	Gstr.	Hls.	Med.	Hrj.	Jmt.	Ång.	Vb.	Nb.	Ås. Lpm.	Ly. Lpm.	P. Lpm.	Lu. Lpm.	T. Lpm.
Embolemus ruddii Westwood	1	●	●		●						●		●		●	●		●	●			●						●		●				●										
Aphelopus melaleucus (Dalman)	2	●	●	●	●				●	●	●		●	●	●	●		●		●		●	●			●		●			●		●			●		●						
A. atratus (Dalman)	3	●	●		●				●		●		●		●	●		●	●			●	●	●		●	●	●	●				●											
A. serratus Richards	4	●	●										●		●	●						●				●		●	●															
A. camus Richards	5	●	●												●			●				●		●		●		●					●											
A. nigriceps Kieffer	6	●	●							●					●								●			●			●															
A. querceus Olmi	7		●																							●			●															
Lonchodryinus ruficornis (Dalman)	8	●	●	●	●	●	●		●	●		●	●		●	●		●	●	●		●	●			●	●	●	●	●	●		●	●	●	●	●	●	●	●		●	●	●
Anteon jurineanum Latreille	9	●	●	●	●				●	●	●		●		●	●		●	●			●	●			●	●	●	●	●			●											
A. brachycerum (Dalman)	10	●	●	●	●							●	●		●	●		●	●			●				●		●				●						●						
A. arcuatum Kieffer	11	●	●		●		●		●				●		●	●						●	●			●		●	●									●						
A. flavicorne (Dalman)	12	●	●	●							●		●		●			●					●					●	●	●														
A. ephippiger (Dalman)	13	●	●	●	●				●				●		●				●	●			●			●		●										●						
A. pubicorne (Dalman)	14	●	●	●	●		●	●	●	●	●		●	●	●	●		●	●	●		●	●	●			●	●		●		●	●	●		●		●	●				●	
A. infectum (Haliday)	15	●	●	●				●	●	●		●			●									●																				
A. exiguum (Haupt)	16	●	●							●					●	●	●					●	●			●	●	●	●	●	●		●											
A. tripartitum Kieffer	17	●	●	●					●	●												●				●		●	●															
A. gaullei Kieffer	18	●	●		●		●		●	●			●		●	●	●	●	●	●		●	●			●	●		●	●														
A. fulviventre (Haliday)	19	●	●	●	●				●	●			●		●	●	●	●				●	●			●		●	●			●												
Dryinus niger Kieffer	20		●	●														●											●	●														
Haplogonatopus oratorius (Westwood)	21		●																																								
Gonatopus helleni (Raatikainen)	22	●							●				●		●	●		●	●			●	●			●	●	●	●															
G. bicolor (Haliday)	23	●	●									●	●		●			●				●				●	●	●					●											
G. pallidus (Ceballos)	24														●																													
G. formicicolus (Richards)	25		●																																									
G. dromedarius (A. Costa)	26														●																													
G. distinctus Kieffer	27	●	●	●												●					●								●															
G. pedestris Dalman	28	●	●						●				●		●								●										●											
G. lunatus Klug	29	●	●					●	●				●		●				●							●	●	●																
G. striatus Kieffer	30	●	●						●			●	●		●	●		●	●	●		●				●		●																
G. spectrum (Van Vollenhoven)	31	●													●		●																											
G. distinguendus Kieffer	32	●	●						●				●		●		●		●	●		●																						
G. formicarius Ljungh	33	●					●	●					●	●	●	●	●	●	●	●						●		●																●
G. clavipes (Thunberg)	34	●	●						●	●	●		●		●	●	●	●		●		●				●	●	●	●		●				●			●		●				
G. albifrons Olmi	35														●																													

Literature

Abdul-Nour, H. 1976. Les Dryininae du sud de la France (Hym., Dryininae). Notes taxonomiques et biologiques; description de deux nouveaux genres. – Annls Zool. Ecol. anim. 8: 265–278.

Arzone, A., Alma, A. & Arnò, C. 1987. Parasitoids and predators of *Rhytidodus decimusquartus* (Rhynchota Auchenorrhyncha). – Proc. 6th Auchenorr. Meeting, Turin, Italy, 7–11 Sept. 1987: 575–580.

Berland, L. 1928. Hyménoptères Vespiformes. II. – Faune Fr. 19, 208 pp. Paris.

Bridwell, J.C. 1958. Biological notes on *Ampulicomorpha confusa* Ashmead and its Fulgoroid host (Hymenoptera: Dryinidae and Homoptera: Achilidae). – Proc. ent. Soc. Wash. 60: 23–26.

Brothers, D.J. & Carpenter, J.M. 1993. Phylogeny of Aculeata: Chrysidoidea and Vespoidea (Hymenoptera). – J. Hymenopt. Res. 2: 227–304.

Brues, C.T. 1923. Some new fossil parasitic Hymenoptera from Baltic Amber. – Proc. Am. Acad. Arts Sci. 58: 327–346.

– 1933. The parasitic Hymenoptera of the Baltic Amber, Part I. Bernstein-forschungen (Amber studies), 3. 178 pp. Berlin und Leipzig.

Burn, J.T. 1993. A host for *Anteon tripartitum* Kieffer (Hym., Dryinidae). – Entomologist's mon. Mag. 129: 65–66.

Buyckx, J.E. 1948. Recherches sur un dryinide, *Aphelopus indivisus*, parasite de Cicadines. – Cellule 52: 63–155.

Carpenter, J.M. 1986. Cladistics of the Chrysidoidea (Hymenoptera). – Jl N. Y. ent. Soc. 94: 303–330.

Ceballos, G. 1927. Anteoninos del Museo de Madrid (Hym. Bethylidae). – Eos, Madr. 3: 97–109.

Chambers, V.H. 1955. Some hosts of *Anteon* spp. (Hym. Dryinidae). – Entomologist's mon. Mag. 91: 114–115.

– A host for *Ismarus halidayi* Foerst. (Hym. Diapriidae). – Ibid. 118: 29.

Chandra, G. 1978. The dryinid parasitoids of the rice leafhoppers and planthoppers in Philippines and techniques of their rearing and mass-rearing. – 9th A. Conf. P.C.C.P. 18 pp. Manila.

– 1980a. Taxonomy and bionomics of rice leafhoppers and planthoppers in the Philippines and their importance in natural biological control. – Philipp. Ent. 4: 119–139.

– 1980b. Dryinid parasitoids of rice leafhoppers and planthoppers in the Philippines. I. Taxonomy and bionomics. – Acta oecol. Appl. 1: 161–172.

Chitty, A.J. 1908. On the Proctotrypid genus *Antaeon*, with descriptions of new species and a table of those occurring in Britain. – Entomologist's mon. Mag. 44: 141–146; 209–215.

Costa, A. 1882. Notizie ed osservazioni sulla geo-fauna sarda. Memoria prima. Risultamento di ricerche fatte in Sardegna nel settembre 1881. – Atti R. Accad. Sci. Napoli 9: 1–41.

Curtis, J. 1828. British Entomology 5. 206 pp. London.

– 1829–30. Guide to an arrangement of British Insects. 110 pp. London.

Dalman, J.W. 1818. Några nya Genera och Species af Insekter beskrifna. – K. svenska VetenskAkad. Handl. 39: 69–89.

– 1823. Analecta Entomologica. 104 pp. Holmiae.

Day, M.C. 1977. A new genus of Plumariidae from Southern Africa, with notes on Scolebythidae (Hymenoptera: Chrysidoidea). – Cimbebasia 4: 171–177.

– 1979. The species of Hymenoptera described by Linnaeus in the genera *Sphex*, *Chrysis*, *Vespa*, *Apis* and *Mutilla*. – Biol. J. Linn. Soc. 12: 45–84.

Delvare, G. & Bouček, Z. 1992. On the new world Chalcididae (Hymenoptera). – Mem. Am. ent. Inst. 53: 1–466.

Dumbleton, L.J. 1937. Apple leaf-hopper investigations. – N. Z. Jl Sci. Technol. 18: 866–877.

Eady, R.D. 1968. Some illustrations of microsculpture in the Hymenoptera. – Proc. R. ent. Soc. Lond. (A) 43: 66–72.

Fenton, F.A. 1918a. The parasites of leaf-hoppers. With special reference to the biology of the Anteoninae. Part I. – Ohio J. Sci. 18: 177–212.

– 1918b. The parasites of leaf-hoppers. With special reference to the biology of the Anteoninae. Part II. – Ibid. 18: 243–278.

– 1918c. The parasites of leaf-hoppers. With special reference to the biology of the Anteoninae. Part III. – Ibid. 18: 285–291.

Finnamore, A.T. & Brothers, D.J. 1993. Superfamily Chrysidoidea. Pp. 130–160 *in* Goulet & Huber: Hymenoptera of the world: an identification guide to families. VII + 668 pp. Ottawa.

Förster, A. 1856. Hymenopterologische Studien, 2. Chalcidiae und Proctotrupii. 152 pp. Aachen.

Forsius, R. 1925. *Embolemus ruddii* Westw. funnen i Finland. – Meddn Soc. Fauna Flora fenn. 49: 61.

Gauld, I. & Bolton, B. 1988. The Hymenoptera. XI + 332 pp. Oxford.

Giri, M.K. & Freytag, P.H. 1989. Development of *Dicondylus americanus* (Hymenoptera: Dryinidae). – Frustula ent., N.S., 9: 215–222.

Gonzalo Abril, R. 1988. Observaciones sobre la biologia del *Tridryinus poecilopterae* (Rich.) (Hymenoptera: Dryinidae) parasito de la Seudopolilla algodonosa *Poekilloptera phalaenoides* (L.) (Homoptera: Flatidae). – Entomólogo 71: 1–5.

Greve, L. & Hauge, E. 1989. Insekt- og edderkopp-faunaen på myrer i Hordaland. 35 pp. Bergen.

Haliday, A.H. 1833. An essay on the classification of the Parasitic Hymenoptera of Britain, which correspond

with the Ichneumones minuti of Linnaeus. - Ent. Mag. 1: 259–273.
- 1838. Note on *Dryinus*. - Ibid. 5: 518.
Hansen, L.O. & Olmi, M. 1994. Aculeata of Norway. 2. Distributional notes on Dryinidae and Embolemidae (Hym., Apocrita). - Fauna norveg., Ser. B, (in press).
Haupt, H. 1932. Die Mundteile der Dryinidae (Hym.). - Zool. Anz. 99: 1–18.
- 1941. Zur Kenntnis der Dryinidae II (Hymenoptera Sphecoidea). - Z. Naturw. 95: 27–67.
Hedqvist, K.-J. 1975. Notes on Embolemidae and Bethylidae in Sweden with description of a new genus and species (Hym., Bethyloidea). - Ent. Tidskr. 96: 121–132.
Heikinheimo, O. 1957. *Dicondylus lindbergi* sp. n. (Hym., Dryinidae), a natural enemy of *Delphacodes pellucida* (F.). - Annls ent. fenn. 23: 77–85.
Hellén, W. 1919a. En för faunan ny representant för parasitstekelslätet *Gonatopus*. - Meddn Soc. Fauna Flora fenn. 45: 185–186.
- 1919b. Zur Kenntnis der Bethyliden und Dryiniden Finlands. - Ibid. 45: 277–290.
- 1930. Inventa entomologica itineris Hispanici et Maroccani, quod a. 1926 fecerunt Harald et Håkan Lindberg. VI. Dryinidae et Bethylidae. - Commentat. biol. 3: 1–6.
- 1935. Dryinidae. Pp. 7–8 *in* Forsius & Hellén: Enumeratio Insectorum Fenniae, II. Hymenoptera. 1. Symphyta et Aculeata. 15 pp. Helsinki.
- 1946. Tre för faunan nya arter av *Gonatopus*. - Notul. ent. 26: 109.
- 1953. Übersicht über die Bethyliden und Dryiniden Finnlands. - Ibid. 33: 88–102.
Hernandez, M.P. 1984. Ciclo de vida y habitos de *Haplogonatopus hernandezae* Olmi (Hymenoptera: Dryinidae) controlador natural del saltahojas del Arroz *Sogatodes oryzicola*. 67 pp. Cali.
- & Belloti, A. 1984. Ciclos de vida y habitos de *Haplogonatopus hernandezae* Olmi (Hymenoptera: Dryinidae) controlador natural del saltahojas del arroz *Sogatodes orizicola* (Muir). - Revta colomb. Ent. 10: 3–8.
Hilpert, H. 1989. Zum Vorkommen einiger Dryiniden in Südwestdeutschland sowie Bemerkungen zu *Embolemus ruddii* Westwood, 1833 (Hymenoptera, Bethyloidea, Dryinidae, Embolemidae). - Spixiana 11: 263–269.
Jansson, A. 1950a. *Anteon (Chelogynus) flaviscapus* nov. sp. (Hymen., Bethylidae). - Ent. Tidskr. 71: 221–222.
- 1950b. Studier över Thomsonska microhymenoptertyper. 1–2. - Opusc. ent. 15: 121–125.
Jervis. M.A. 1977. A new key for the identification of the British species of *Aphelopus* (Hym.: Dryinidae). - Syst. Ent. 2: 301–303.
- 1979a. Parasitism of *Aphelopus* species (Hymenoptera: Dryinidae) by *Ismarus dorsiger* (Curtis) (Hymenoptera: Diapriidae). - Entomologist's Gaz. 30: 127–129.
- 1979b. Courtship, mating and 'swarming' in *Aphelopus melaleucus* (Dalman) (Hymenoptera: Dryinidae). - Ibid. 30: 191–193.
- 1980a. Studies on oviposition behaviour and larval development in species of *Chalarus* (Diptera: Pipunculidae), parasites of typhlocybine leafhoppers (Homoptera, Cicadellidae). - J. nat. Hist. 14: 759–768.
- 1980b. Life history studies on *Aphelopus* species (Hymenoptera, Dryinidae) and *Chalarus* species (Diptera, Pipunculidae), primary parasites of typhlocybine leafhoppers (Homoptera, Cicadellidae). - Ibid. 14: 769–780.
- 1980c. Ecological studies on the parasitic complex associated with typhlocybine leafhoppers (Homoptera, Cicadellidae). - Ecol. Ent. 5: 123–136.
- 1986. New host records for *Aphelopus* (Hymenoptera: Dryinidae). - Entomologist's Gaz. 37: 37–38.
- & Kidd, N.A.C. 1986. Host-feeding strategies in Hymenopteran parasitoids. - Biol. Rev. 61: 395–434.
- Kidd, N.A.C. & Sahragard, A. 1987. Host-feeding in Dryinidae: its adaptative significance and its consequences for parasitoid-host population dynamics. - Proc. 6th Auchenorr. Meeting, Turin, Italy, 7–11 Sept. 1987: 591–596.
Jurine, L. 1807. Nouvelle méthode de classer les Hyménoptères et les Diptères. 1. 320 + 4 pp. Genève.
Keilin, D. & Thompson, W.R. 1915. Sur le cycle évolutif des Dryinidae, Hyménoptères parasites des Hémiptères Homoptères. - C. r. Séanc. Soc. Biol. 1915: 5–9.
Kieffer, J.-J. 1904. Description de nouveaux Dryininae et Bethylinae du Musée Civique de Gènes. - Annali Mus. civ. Stor. nat. Genova 41: 351–412.
- 1905. Description de nouveaux Proctotrypidae exotiques. - Annls Soc. scient. Brux. 29: 95–142.
- 1907. Fam. Dryinidae. - Genera Insect., 54. 33 pp., Bruxelles.
- 1911. Description d'un nouveau dryinide des Indes orientales. - Bull. Soc. Hist. nat. Metz 27: 107–110.
- 1913. Division des Anteoninae (Hym.). - Bull. Soc. ent. Fr. 1913: 300–301.
- 1914. Bethylidae. - Tierreich, 41. 395 pp. Berlin.
- & Marshall, T.A. 1904–1906. Proctotrypidae. Pp. 1–552 *in* André: Species des Hyménoptères d'Europe et d'Algérie, 9. 552 pp. Paris.
Kitamura, K. 1982. Comparative studies on the biology of dryinid wasps in Japan (1). Preliminary report on the predacious and parasitic efficiency of *Haplogonatopus atratus* Esaki at Hashimoto (Hymenoptera: Dryinidae). - Bull. Fac. Agric. Shimane Univ. 16: 172–176.
- 1983. Comparative studies on the biology of dryinid wasps in Japan (2). Relationships between temperature and the developmental velocity of *Haplogonatopus atratus* Esaki et Hashimoto (Hymenoptera: Dryinidae). - Ibid. 17: 147–151.
- 1985. Comparative studies on the biology of dryinid wasps in Japan (3). Immature stages and morphology of *Haplogonatopus atratus* Esaki et Hashimoto (Hymenoptera: Dryinidae). - Ibid. 19: 154–158.
- 1986. Comparative studies on the biology of dryinid wasps in Japan (4). Longevity, oviposition and host-feeding of adult female of *Haplogonatopus atratus* Esaki et Hashimoto (Hymenoptera: Dryinidae). - Ibid. 20: 191–195.
- 1988. Comparative studies on the biology of dryinid wasps in Japan (5). Development and reproductive capacity of hosts attacked by *Haplogonatopus apicalis*

(Hymenoptera, Dryinidae) and the development of progenies of the parasites in their hosts. - Kontyû 56: 659–666.
- 1989a. Comparative studies on the biology of dryinid wasps in Japan (6). Hibernation and development of *Haplogonatopus atratus* Esaki et Hashimoto (Hymenoptera: Dryinidae) on overwintering leaf- and planthoppers (Homoptera: Auchenorrhyncha). - Jap. J. appl. Ent. Zool. 33: 24–30.
- 1989b. Comparative studies on the biology of dryinid wasps in Japan (VIII). The daily periodicity of oviposition and predation of *Haplogonatopus atratus* Esaki et Hashimoto (Hymenoptera: Dryinidae). - Ibid. 33: 140–141.
- & Nishikata, Y. 1987. A monitor-trap survey of parasitoids of the leaf- and planthoppers supposedly migrated from the mainland China (Homoptera: Auchenorrhyncha). - Bull. Fac. Agric. Shimane Univ. 21: 171–177.
Klug, J.C.F. 1810. Ueber die Ljunghsche Piezaten-gattung *Gonatopus*. - Beitr. Naturk. 2: 164–165.
Kontkanen, P. 1950. Notes on the parasites of leafhoppers in North Karelia. - Annls ent. fenn. 16: 101–109.
Krogerus, R. 1932. Über die Ökologie und Verbreitung der Arthropoden der Triebsandgebiete an den Küsten Finnlands. - Acta zool. fenn. 12: 1–308.
Latreille, P.A. 1804. Nouvelle dictionnaire d'Histoire naturelle, 24. Paris.
- 1809. Genera crustaceorum et insectorum secundum ordinem naturalem in familias disposita, iconibus exemplisque plurimis explicata, 4. 399 pp. Paris and Strasbourg.
Le Quesne, W.J. 1972. Studies on the coexistence of three species of *Eupteryx* (Hemiptera: Cicadellidae) on nettle. - J. Ent. (A) 47: 37–44.
Lindberg, H. 1950. Notes on the biology of dryinids. - Commentat. biol. 10: 1–19.
Linnaeus, C. 1767. Systema Naturae per regna tria naturae, secundum classes, ordines, genera, species, cum caracteribus, differentiis, synonymis, locis. 12th edition (revised), 1 (2). Pp. 533–1327. Holmiae.
Ljungh, S.J. 1810. *Gonatopus*, novum insectorum genus. - Beitr. Naturk. 2: 161–163.
Masner, L. 1976. A revision of the Ismarinae of the New World (Hymenoptera, Proctotrupoidea, Diapriidae). - Can. Ent. 108: 1243–1266.
Moczar. L. 1967. On Kieffer's and Dalman's Types (Hymenoptera, Dryinidae). - Annls hist.-nat. Mus. natn. Hung. 59: 297–302.
Naumann, I.D. 1991. Hymenoptera. Pp. 916–1000 *in* CSIRO: The insects of Australia. XIII + 1137 pp. Carlton and London.
Nilsson, G.E. 1986. Nya landskapsfynd av gaddsteklar, med en översikt av de fennoskandiska arterna i familjen Dryinidae. - Ent. Tidskr. 107: 85–90.
- 1988. Nya landskapsfynd av gaddsteklar med *Evagetes subnudus* ny för Nordeuropa och *Sphecodes albilabris* återfunnen i Sverige. - Ibid. 109: 97–100.
- 1991. The wasp and bee fauna of the Ridö archipelago in Lake Mälaren, Sweden (Hymenoptera, Aculeata). - Ibid. 112: 79–92.
Nixon, G.E.J. 1957. Hymenoptera Proctotrupoidea Diapriidae, subfamily Belytinae. - Handbk Ident. Br. Insects VIII (3d). 107 pp. London.
Olmi, M. 1977. Revision of the C.G. Thomson species of Gonatopodinae (Hymenoptera: Dryinidae). - Ent. scand. 8: 157–158.
- 1984. A revision of the Dryinidae (Hymenoptera, Chrysidoidea). - Mem. Am. ent. Inst. 37. XXXI + 1913 pp. Ann Arbor.
- 1987a. New species of Dryinidae (Hymenoptera, Chrysidoidea). - Fragm. ent. 19: 371–456.
- 1987b. New species of Dryinidae, with description of a new subfamily from Florida and a new species from Dominica amber (Hymenoptera, Chrysidoidea). - Boll. Mus. R. Sci. nat. Torino 5: 211–238.
- 1989. Supplement to the revision of the world Dryinidae (Hymenoptera Chrysidoidea). - Frustula ent., N.S., 12: 109–395.
- 1993. A new generic classification for Thaumatodryininae, Dryininae and Gonatopodinae, with descriptions of new species (Hymenoptera Dryinidae). - Boll. Zool. agr. Bachic., Ser. II, 25: 57–89.
- 1994. A revision of the world Embolemidae (Hymenoptera Chrysidoidea). - Frustula ent. (in press).
- & Currado, I. 1977. On the identity of *Gonatopus pedestris* Dalman (Hymenoptera: Dryinidae). - Ent. scand. 8: 76–78.
Ossiannilsson, F. 1978. The Auchenorrhyncha (Homoptera) of Fennoscandia and Denmark. Part 1: Introduction, infraorder Fulgoromorpha. - Fauna ent. scand. 7 (1): 1–222.
- 1981. The Auchenorrhyncha (Homoptera) of Fennoscandia and Denmark. Part 2: The families Cicadidae, Cercopidae, Membracidae, and Cicadellidae (excl. Deltocephalinae). - Ibid. 7 (2): 223–593.
- 1983. The Auchenorrhyncha (Homoptera) of Fennoscandia and Denmark. Part 3: The Family Cicadellidae: Deltocephalinae, Catalogue, Literature and Index. - Ibid. 7 (3): 594–979.
Pagden, H. 1934. Notes on hymenopterous parasites of padi insects in Malaya. - Scient. Ser. Dep. Agric. Fed. Malaya 15: 13 pp. Kuala Lumpur.
Perkins, J.F. 1976. Hymenoptera Bethyloidea. - Handbk Ident. Br. Insects VI (3a). 38 pp. London.
Perkins, R.C.L. 1903. The leafhopper of the sugar cane. - Bull. Terr. Hawaii Bd agric. for. Ent. 1: 1–38.
- 1905. Leafhoppers and their natural enemies (Pt. I. Dryinidae). - Bull. Hawaiian Sug. Plrs' Ass. Exp. Stn, Ent., 1 (1): 1–69.
- 1906a. Leafhoppers and their natural enemies (Pt. VIII. Encyrtidae, Eulophidae, Trichogrammidae). - Ibid., Ent., 1 (8): 241–267.
- 1906b. Leafhoppers and their natural enemies (Pt. X. Dryinidae, Pipunculidae). - Ibid., Ent., 1 (10): 483–499.
- 1906c. Leafhoppers and their natural enemies. - Ibid., Ent., 1: I–XXXII.
- 1907. Parasites of leafhoppers. - Ibid., Ent., 2 (4): 5–59.
- 1912. Parasites of the family Dryinidae. - Ibid., Ent., 3 (11): 5–20.
Pillault, R. 1951. Notes sur *Dryinus tarraconensis* (Hym. Dryinidae) prédateur d'une cicadelle. - Annls Soc. ent. Fr. 120: 67–76.

Ponomarenko, N.G. 1971. Some peculiarities of development of Dryinidae. – Proc. XIII Int. Congr. Ent. Moscow 1968, 1: 281–282.
– 1975. Characteristics of larval development in the Dryinidae (Hymenoptera). – Ent. Obozr. 54: 534–540.
– 1978. Dryinidae. Pp. 16–27 *in* Medvedev: Key to the insects of the European part of the USSR, 3, Hymenoptera. 1341 pp. Moscow.
Poinar Jr, G.O. 1993. Insects in amber. – Ann. Rev. Ent. 46: 145–159.
Raatikainen, M. 1960. The biology of *Calligypona sordidula* (Stål) (Hom., Auchenorrhyncha). – Annls ent. fenn. 26: 229–242.
– 1961. *Dicondylus helléni* n. sp. (Hym. Dryinidae), a parasite of *Calligypona sordidula* (Stål) and *C. excisa* (Mel.). – Ibid. 27: 126–137.
– 1967. Bionomics, enemies and population dynamics of *Javesella pellucida* (F.) (Hom. Delphacidae). – Annls agric. fenn. 6 (Suppl. 2): 1–149.
– 1970. Ecology and fluctuations in abundance of *Megadelphax sordidula* (Stål) (Hom. Delphacidae). – Ibid. 6: 315–324.
– & Vasarainen, A. 1964. Biology of *Dicranotropis hamata* (Boh.) (Hom. Araeopidae). – Ibid. 3: 311–323.
Richards, O.W. 1939. The British Bethylidae (s.l.) (Hymenoptera). – Trans. R. ent. Soc. Lond. 89: 185–344.
Sahlberg, J. 1910. Om parasitstekel-slägtet *Gonatopus* och dess Finska representanter (Hym.). – Acta Soc. Fauna Flora fenn. 33: 118–125.
Strand, E. 1898. Enumeratio Hymenopterorum Norvegicorum. – Ent. Tidskr. 19: 71–112.
Subba Rao, B.R. 1957. The biology and bionomics of *Lestodryinus pyrillae* Kieff. (Dryinidae: Hymenoptera) a nymphal parasite of *Pyrilla perpusilla* Walk. and a note on its role in the control of *Pyrilla*. – J. Bombay nat. Hist. Soc. 54: 741–749.
Swezey, O.H. 1908. On peculiar deviations from uniformity of habit among Chalcids and Proctotrupids. – Proc. Hawaii. ent. Soc. 4: 2–3.
– 1928. Present status of certain insect pests under biological control in Hawaii. – J. econ. Ent. 21: 669–676.
Teodorescu, I. 1982. Some aspects of the biology of *Gonatopus sepsoides* Westwood (Hym. Dryinidae). – Anuar Univ. Buc., Biol., 31: 67–72.
Thomson, C.G. 1860. Sveriges Proctotruper. – Öfvers. K. VetenskAkad. Förh. Stockh. 17: 169–181.
Thunberg, C.P. 1827. *Gelis* insecti genus descriptum. – Nova Acta R. Soc. Scient. upsal. 9: 199–204.
Tryapitsyn, V.A. 1978. Family Encyrtidae. Pp. 427–594 *in* Medvedev: Keys to the insects of the European part of the USSR, 3, Hymenoptera. 1341 pp. Moscow.
Van Vollenhoven, S.C. 1874. Over de groep der dryiniden in de familie der Proctotrupiden, met beschrijving eener nieuwe soort. – Versl. Meded. K. Akad. wet. Amst., Afd. Naturk., (2) 8: 150–162.
Vikberg, V. 1986. A checklist of aculeate Hymenoptera of Finland (Hymenoptera, Apocrita Aculeata). – Notul. ent. 66: 65–85.
Walker, F. 1837. On the Dryinidae. – Ent. Mag. 4: 411–435.
Waloff, N. 1974. Biology and behaviour of some species of Dryinidae (Hymenoptera). – J. Ent. (A) 49: 97–109.
– 1975. The parasitoids of the nymphal and adult stages of leafhoppers (Auchenorrhyncha: Homoptera) of acidic grassland. – Trans. R. ent. Soc. Lond. 126: 637–686.
– 1990. Superparasitism and multiparasitism of Cicadellidae and Delphacidae (Homoptera, Auchenorrhyncha). – Entomologist 109: 47–52.
– & Jervis, M.A. 1987. Communities of parasitoids associated with leafhoppers and planthoppers in Europe. – Adv. ecol. Res. 17: 281–402.
– & Thompson, P. 1980. Census data of populations of some leafhoppers (Auchenorrhyncha, Homoptera) of acidic grassland. – J. anim. Ecol. 49: 395–416.
Westwood, J.O. 1833a. Descriptions of several new British forms amongst the parasitic hymenopterous insects. – Lond. Edinb. Phil. Mag. 2: 443–445.
– 1833b. Notice of the habits of a Cynipideous Insect, parasitic upon the Rose Louse (*Aphis rosae*); with descriptions of several other parasitic Hymenoptera. – Mag. nat. Hist. 6: 491–497.
– 1835. Genus *Campylonyx*. – Proc. zool. Soc. Lond. 3: 52.
Wharton, R.A. 1989. Final instar larva of the Embolemid wasp *Ampulicomorpha confusa* (Hymenoptera). – Proc. ent. Soc. Wash. 91: 509–512.
Williams, F.X. 1931. Handbook of the insects and other invertebrates of Hawaiian sugar cane fields. 400 pp. Honolulu.
Wilson, M.R. & Claridge, M.F. 1991. Handbook for the identification of leafhoppers and planthoppers of rice. 142 pp. Oxon.

Index

Synonyms are given in italics. Numbers in bold refer to the main treatment of each taxon.

FAUNA ENTOMOLOGICA SCANDINAVIA

EDITED BY

N.P. KRISTENSEN AND V. MICHELSEN

1. Rozkošný, R. *The Stratiomyioidea (Diptera) of Fennoscandia and Denmark.* 1973. ISBN 87 87 49100 1
2. Fibiger, M.; Kristensen, N.P. *The Sesiidae (Lepidoptera) of Fennoscandia and Denmark.* 1974. ISBN 87 87 49102 8
3. Chvála, M. *The Tachydromiinae (Dipt. Empididae) of Fennoscandia and Denmark.* 1975. ISBN 87 87 49104 4
4. Lomholdt, O. *The Sphecidae (Hymenoptera) of Fennoscandia and Denmark.* 2nd ed. 1984. ISBN 87 87 49106 0
5. Spencer, K.A. *The Agromyzidae (Diptera) of Fennoscandia and Denmark.* 1976. ISBN 87 87 49108 7
6. Traugott-Olsen, E.; Nielsen, E.S. *The Elachistidae (Lepidoptera) of Fennoscandia and Denmark.* 1977. ISBN 87 87 49114 1

7/1. Ossiannilsson, F. *The Auchenorrhyncha (Homoptera) of Fennoscandia and Denmark.* 1. Introduction, infraorder Fulgoromorpha. 1978. ISBN 87 87 49124 9

7/2. Ossiannilsson, F. *The Auchenorrhyncha (Homoptera) of Fennoscandia and Denmark.* 2. The families Cicadidae, Cercopidae, Membracidae and Cicadellidae (excl. Deltocephalinae). 1981. ISBN 87 87 49136 2

7/3. Ossiannilsson, F. *The Auchenorrhyncha (Homoptera) of Fennoscandia and Denmark.* 3. The family Cicadellidae: Deltocephalinae, catalogue, literature and index. 1983. 87 87 49113 3

8. Collingwood, C. *The Formicidae (Hymenoptera) of Fennoscandia and Denmark.* 1979. ISBN 87 87 49128 1
9. Heie, O. *The Aphidoidea (Hemiptera) of Fennoscandia and Denmark.* 1. General part. The families Mindaridae, Hormaphididae, Thelaxidae, Anoeciidae, and Pemphigidae. 1980. 87 87 49134 6
10. Bilý, S. *The Buprestidae (Coleoptera) of Fennoscandia and Denmark.* 1982. ISBN 87 87 49142 7
11. Heie, O. *The Aphidoidea (Hemiptera) of Fennoscandia and Denmark.* 2. The family Drepanosiphidae. 1982. ISBN 87 87 49144 3
12. Chvála, M. *The Empidoidea (Diptera) of Fennoscandia and Denmark.* II. General part. The families Hybotidae, Atelestidae and Microphoridae. 1983. ISBN 87 87 49107 9
13. Bengtsson, B. *The Scythrididae (Lepidoptera) of Northern Europe.* 1984. ISBN 90 04 07312 4
14. Rozkošný, R. *The Sciomyzidae (Diptera) of Fennoscandia and Denmark.* 1984. ISBN 90 04 07592 5

15/1. Lindroth, C.H. *The Carabidae (Coleoptera) of Fennoscandia and Denmark.* 1. 1985. ISBN 90 04 07727 8

15/2. Lindroth, C.H. *The Carabidae (Coleoptera) of Fennoscandia and Denmark.* 2. With an Appendix on the Family Rhysodidae. 1986. ISBN 90 04 08182 8

16. Holst, K.Th. *The Saltatoria (Bush-crickets, Crickets and Grass-hoppers) of Northern Europe.* 1986. ISBN 90 04 07860 6
17. Heie, O. *The Aphidoidea (Hemiptera) of Fennoscandia and Denmark.* 3. Family Aphididae: Subfamily Pterocommatinae and Tribe Aphidini of Subfamily Aphidinae. 1986. ISBN 90 04 08088 0
18. Hansen, M. *The Hydrophiloidea (Coleoptera) of Fennoscandia and Denmark.* 1987. ISBN 90 04 08183 6
19. Pape, Th. *The Sarcophagidae (Diptera) of Fennoscandia and Denmark.* 1987. ISBN 90 04 08184 4
20. Holmen, M. *The Aquatic Adephaga (Coleoptera) of Fennoscandia and Denmark.* I. Gyrinidae, Haliplidae, Hygrobiidae and Noteridae. 1987. ISBN 90 04 08185 2
21. Lillehammer, A. *Stoneflies (Plecoptera) of Fennoscandia and Denmark.* 1988. ISBN 90 04 08695 1
22. Bilý, S.; Mehl, O. *Longhorn beetles (Coleoptera, Cerambycidae) of Fennoscandia and Denmark.* 1989. ISBN 90 04 08697 8
23. Johansson, R.; Nielsen, E.S.; Nieukerken, E.J. van; Gustafsson, B. *The Nepticulidae and Opostegidae (Lepidoptera) of North and West Europe.* 1991. ISBN 90 04 08698 6
24. Rognes, K. *Blowflies (Diptera, Calliphoridae) of Fennoscandia and Denmark.* 1991. ISBN 90 04 09304 4
25. Heie, O.E. *The Aphidoidea (Hemiptera) of Fennoscandia and Denmark.* IV. Family Aphididae: Part 1 of Tribe Macrosiphini of Subfamily Aphidinae. 1991. ISBN 90 04 09514 4
26. Ossiannilsson, F. *The Psylloidea (Homoptera) of Fennoscandia and Denmark.* 1992. ISBN 90 04 09610 8

27. Luff, M.L. *The Carabidae (Coleoptera) larvae of Fennoscandia and Denmark*. 1993. ISBN 90 04 09836 4
28. Heie, O.E. *The Aphidoidea (Hemiptera) of Fennoscandia and Denmark*. V. Family Aphididae: Part 2 of tribe Macrosiphini of Subfamily Aphidinae. 1993. ISBN 90 04 09899 22
29. Chvála, M. *The Empidoidea (Diptera) of Fennoscandia and Denmark*. III. Genus *Empis*. 1994. ISBN 90 04 09663 9
30. Olmi, M. *The Dryinidae and Embolemidae (Hymenoptera: Chrysidoidea) of Fennoscandia and Denmark*. 1994. ISBN 90 04 10224 8

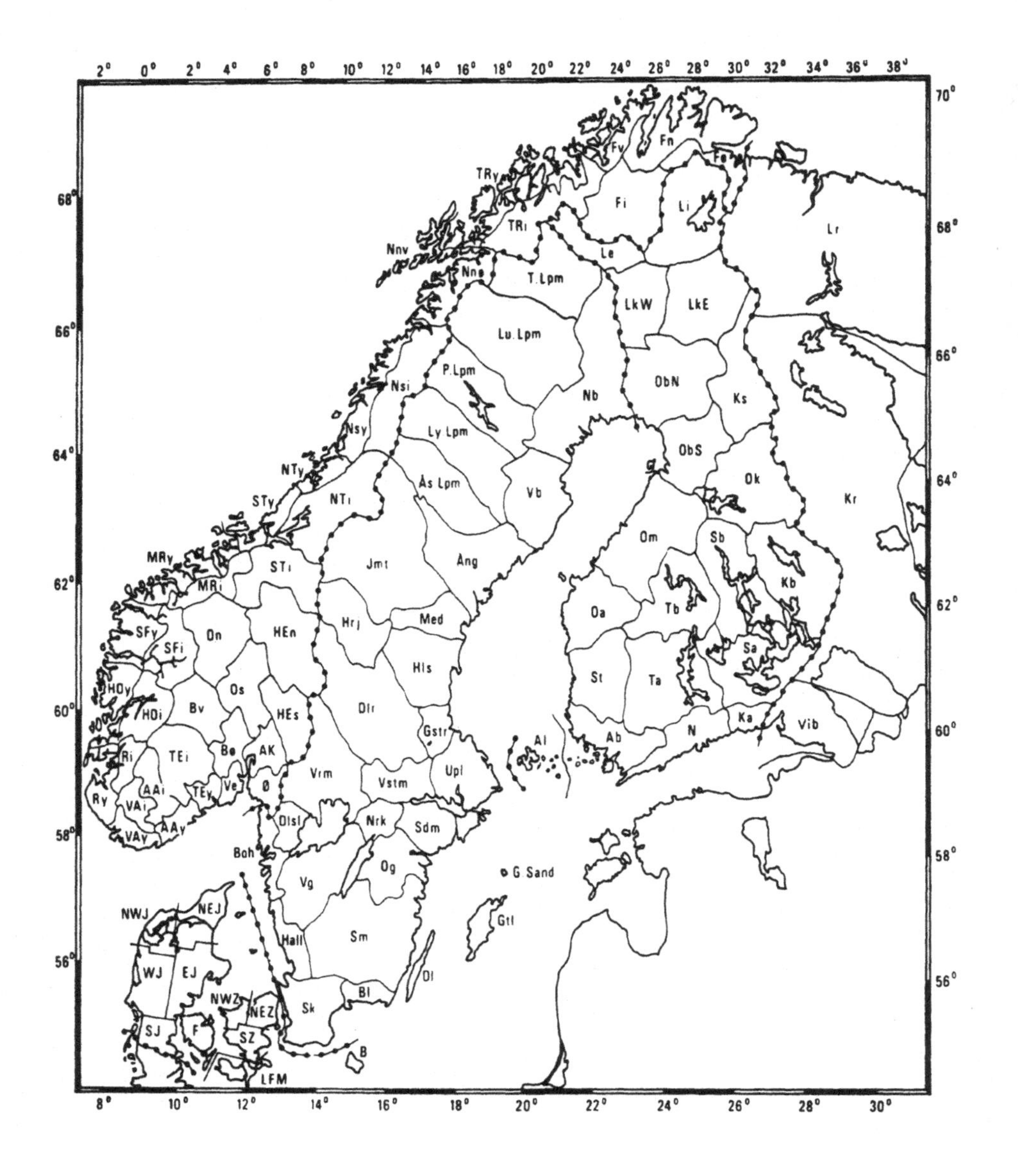

List of abbreviations for the provinces used throughout the text, on the map and in the following tables.

DENMARK

SJ	South Jutland	LFM	Lolland, Falster, Møn
EJ	East Jutland	SZ	South Zealand
WJ	West Jutland	NWZ	North West Zealand
NWJ	North West Jutland	NEZ	North East Zealand
NEJ	North East Jutland	B	Bornholm
F	Funen		

SWEDEN

Sk.	Skåne
Bl.	Blekinge
Hall.	Halland
Sm.	Småland
Öl.	Öland
Gtl.	Gotland
G. Sand.	Gotska Sandön
Ög.	Östergötland
Vg.	Västergötland
Boh.	Bohuslän
Dlsl.	Dalsland
Nrk.	Närke
Sdm.	Södermanland
Upl.	Uppland
Vstm.	Västmanland
Vrm.	Värmland
Dlr.	Dalarna
Gstr.	Gästrikland
Hls.	Hälsingland
Med.	Medelpad
Hrj.	Härjedalen
Jmt.	Jämtland
Ång.	Ångermanland
Vb.	Västerbotten
Nb.	Norrbotten
Ås. Lpm.	Åsele Lappmark
Ly. Lpm.	Lycksele Lappmark
P. Lpm.	Pite Lappmark
Lu. Lpm.	Lule Lappmark
T. Lpm.	Torne Lappmark

NORWAY

Ø	Østfold
AK	Akershus
HE	Hedmark
O	Oppland
B	Buskerud
VE	Vestfold
TE	Telemark
AA	Aust-Agder
VA	Vest-Agder
R	Rogaland
HO	Hordaland
SF	Sogn og Fjordane
MR	Møre og Romsdal
ST	Sør-Trøndelag
NT	Nord-Trøndelag
Ns	southern Nordland
Nn	northern Nordland
TR	Troms
F	Finnmark

n northern s southern ø eastern v western y outer i inner

FINLAND

Al	Alandia
Ab	Regio aboensis
N	Nylandia
Ka	Karelia australis
St	Satakunta
Ta	Tavastia australis
Sa	Savonia australis
Oa	Ostrobottnia australis
Tb	Tavastia borealis
Sb	Savonia borealis
Kb	Karelia borealis
Om	Ostrobottnia media
Ok	Ostrobottnia kajanensis
ObS	Ostrobottnia borealis, S part
ObN	Ostrobottnia borealis, N part
Ks	Kuusamo
LkW	Lapponia kemensis, W part
LkE	Lapponia kemensis, E part
Li	Lapponia inarensis
Le	Lapponia enontekiensis

USSR

Vib	Regio Viburgensis
Kr	Karelia rossica
Lr	Lapponia rossica

Printed in the United States
By Bookmasters